West to East: A New Global Economy in the Making?

Arno J. van Niekerk

West to East: A New Global Economy in the Making?

Achieving the SDGs

Springer

Arno J. van Niekerk
Department of Economics and Finance
University of the Free State
Bloemfontein, South Africa

ISBN 978-3-031-93266-3 ISBN 978-3-031-93267-0 (eBook)
https://doi.org/10.1007/978-3-031-93267-0

This work was supported by Central Research Unit, University of the Free State.

This Springer imprint is published by the registered company Springer Nature Switzerland AG
The registered company address is: Gewerbestrasse 11, 6330 Cham, Switzerland

I dedicate this book to my family for their unwavering support, to my colleagues and students at the University of the Free State (South Africa), and to the people of Africa who are working towards a new and brighter future.

Prologue

At a time of rising global tension, especially between East and West, the global economy is facing growing uncertainty as it navigates uncharted territory in the Fourth Industrial Revolution (4IR). The emergence of new forms of innovation and new powers in the East—most notably China, India, Russia, and several Arab states in the Middle East—is reshaping the global landscape, asking new questions about our evolving economic reality. Meanwhile, traditional power centres in the West—the United States (US) and Europe—appear less focused on offering a *new vision* for specifically the global economy compared to their Eastern counterparts. The US, now led by Donald Trump, follows a very self-centred, inward, 'America First' approach, subsequent to Joe Biden's slow, lethargic and reluctant approach. Europe, while highly concerned about Russia's invasion of Ukraine and its after-effects, is experiencing internal struggles and does not seem to be in a 'creative space' to offer new vision to the world. China, on the other hand, has clear objectives for the world economy, such as building infrastructure and stimulating trade across Asia, Africa, the Middle East, and Latin America as part of its Belt and Road Initiative (BRI); it is promoting multilateralism in response to rising protectionism and trade barriers in other parts of the world; and it is providing progressive technological leadership. Many developing countries are showing growing interest to be part of this *new vision*, which resulted in the expansion of BRICS (Brazil, Russia, India, China and South Africa) to BRICS+ (adding Egypt, Iran, UAE and Ethiopia so far). Several other countries are lining up to join the momentum surging from the East.

Another 4IR development promising to have a significant impact on the evolution of the global economy in the 21st century is the digital revolution in especially finance, which encompass the rise of cryptocurrencies, the introduction of central bank digital currencies (CBDCs), and the proliferation of FinTech and artificial intelligence (AI). This revolution is sketching a radically *new vision* for how the global economy can function in the next decade. In light of these transformations, an area that is drawing particular attention is the role of the Sustainable Development Goals (SDGs), which aim to guide economies toward greater sustainability. Two perspectives have emerged: some question how realistic achieving the SDGs are after the COVID-19 pandemic, while others advocate for their increasing significance. As we approach 2030, it is vital to ask whether the SDGs should still be a primary priority for global progress or have tech innovation and heightened competition between East and West taken precedence in determining global progress? The answer is arguably that all these dynamics are not mutually exclusive, thus re-emphasizing the importance of the SDGs as a balancing factor, a 'bulls eye' to keep aiming at. The concern that looms large is the potential Prisoner's Dilemma in the global economy, where rising tensions between individual self-interest and collective benefit may lead rational leaders to avoid cooperation, even when it serves their best interests. Or could existential necessity instead drive global collaboration, especially through a shared aim like economic inclusion, which is vital to achieving the SDGs? These are critical developments that will be investigated in this book.

Without a doubt, the option and possibility of building bridges and foster more constructive global cooperation—as opposed to increased hostility—is better. For the sake of human and planetary sustainability, economic inclusion can truly become the unifying factor that cause both East and West to choose a higher path, a *grander vision*, for a type of global prosperity that benefits all. The future of the world, with the economy as the decisive element, is now at a crucial crossroads. A path of self-centredness risks deepening fractures and mutual destruction in the global economy, while one of greater collaboration between East and West promises genuine progress for all. In simple terms, the big question for humanity is what is more important, teamwork or self-interest?

No one can deny that the momentum in the global economy has changed. The East is growing stronger and the West does not hold the same power over world affairs like between World War II and the 2007–2009 global financial crisis (GFC) or perhaps the COVID-19 pandemic. Ever since, the West's economic influence over the world is visibly declining, which raises the question of whether the US and the European Union (EU) will respond defensively, feeling threatened. On the other hand, the rise of China, in particular, having become a global giant in merely 30 years, has changed the dynamics of geopolitics. It seems like China's alliances and rivalries are increasingly shaping the course of global political and economic developments. Confirming this, the former leader of the World Economic Forum (WEF), Klaus Schwab, explicitly stated that China is a "role model" for many countries (Billingsley, 2022). A major question is whether China and the East will have the necessary restraint to avoid abusing or exploiting their growing influence in world affairs? This area of concern is a central focus in the book. It investigates the tension between the possible, comparative declining influence of the West and the surging global influence of the East in view of a technology revolution that is becoming a competitive undercurrent of change in both trends.

Key questions immediately arise like what are the implications of these major global developments for economic sustainability? Are we seeing a new form of capitalism emerging or is it new socialism? If there is a move towards de-dollarization in the East, what are their intentions and what are the consequences for the global economy? Do the SDGs remain a high priority for all countries, or is progress now driven solely by self-interest? What is the role of global institutions (in the West and East) regarding globalization and/or de-globalization? The book investigates these questions to identify a pathway towards sustainable global economic progress. If economic inclusion develops into a central objective, a greater degree of international collaboration will be essential to steering the world economy toward meaningful and shared prosperity. This is a more complete *vision* at a higher level of thinking than mere self-interest. The book's objective is to examine these key global trajectories and dynamics that are shaping a new world and to propose a way forward. The geopolitical and geoeconomic shifts are carving a new international economic order of competing economic blocs in the East and West. This requires thorough investigation.

The volume synthesizes diverse themes in this exploration to reach the objective mentioned above. Its distinctiveness is shown by several timely contributions to literature, seeking to inform policymakers, leaders, and motivate scholars to do research in these dynamic areas. These include: comparative trends of key geopolitical and geoeconomic shifts over the past 30 years; illustrations of how the global financial system is evolving; new analyses of traditional economic theories (e.g., on capitalism and socialism) in the 21st century context; the components of an inclusive economy and how they can fast-track countries towards SDG-attainment; new interpretations of the digital revolution in the context of geoeconomic transitions; and new proposals on how to deal with the challenges related to the aspects mentioned from a global economic governance perspective. This volume combines academic analysis and practical examples in a meaningful way to improve economic decision-making at policy level, in business, and at community level. The hope is that it inspires experts and practitioners to further investigate and apply the insights generated in this book.

Bloemfontein, South Africa Arno J. van Niekerk

Reference

Billingsley, K. L. (2022). Klaus Schwab proclaims China a "role model" for many nations. https://blog.independent.org/2022/12/01/klaus-schwab-china/.

Preface

This book seeks to navigate the intricate and rapidly evolving landscape of the global economy, especially in light of a visible geopolitical and geoeconomic shift from West to East. The global economic landscape is changing in profound and unprecedented ways with the rise of China, India, and BRICS, as well as groundbreaking technological innovations. The world is witnessing the emergence of new economic paradigms, trade alliances, and financial systems that challenge the long-standing dominance of Western economies. This volume seeks to examine the forces driving this transformation, offering a comprehensive analysis of the geopolitical, economic, and social dynamics that are reshaping our world. The scope of this exploration is broad, encompassing the economic implications of the West-to-East geopolitical transition, the sustainability challenges posed by a comparatively declining US economy and the movement towards de-dollarization, and the transformative role of the digital revolution in finance and its impact on achieving the Sustainable Development Goals (SDGs).

The motivation for writing this book stems from an increasing need to understand the implications of these changes and a deep concern for the future of the global economy, and a conviction that fostering constructive global cooperation is paramount for human and planetary sustainability. Observing the changing momentum in the global economy, the rapid rise of China as a global economic power stands out as a predominant factor in new geopolitical dynamics. Traditional economic models and institutions, primarily shaped by Western ideologies, are being tested as Eastern emerging economies, in particular, assert their influence. They are not only growing at remarkable rates but are also redefining global economic governance, trade partnerships, and financial stability mechanisms. At the same time, technological advancements, environmental crises, and socio-economic disparities are accelerating the need for new strategies in global economic management. Understanding these shifts is crucial for developing new insights that can inform policymakers, business and political leaders, scholars, and individuals who seek to navigate the complexities of the 21st-century global economy. This book is driven by the need to understand the implications of these shifts for economic sustainability, to analyze the potential for new economic models, to assess the future of the dollar's dominance, and to determine the continued relevance of the SDGs in this evolving context.

Beyond economic realignments, the book addresses the broader socio-political implications of this transformation. What does the rise of the East mean for democracy, human rights, and governance models? How are international institutions adapting to the changing balance of power? What new challenges and opportunities arise from these shifts, and how can nations cooperate to ensure sustainable and inclusive economic growth? One of the central themes of this book is the need for economic inclusion and collaboration. The new global economy must not replicate the inequalities and exploitative practices of the past. Instead, it should aim for a more equitable distribution of resources, fostering development that benefits all, rather than a select few. By understanding the mechanisms of economic power shifts, stakeholders can make informed decisions to shape a future that prioritizes both prosperity and social justice.

This book is not just an academic analysis; it is a call to action. It urges policymakers, leaders, and global citizens to engage with these changes proactively, fostering a global economy that is dynamic, fair, and sustainable. In an era of rapid technological advancements, climate challenges, and increasing interdependence, the world must embrace new economic models that reflect the realities of the 21st century. This would inform and encourage a move towards a more collaborative and inclusive global economic future that benefits all.

This volume is structured to provide a balanced and critical perspective on these developments. It begins in Chap. 1 by examining the economic implications of the probable geopolitical shift from West to East. Chapter 2 explores the challenges that a comparatively declining US economy, de-dollarization and the rise of the East present to making progress towards economic sustainability, globally. Chapter 3 investigates the role of the digital revolution in especially the area of finance for gaining a competitive edge (for global dominance), and what its implications are for countries to attain the SDGs. Chapter 4 delves into the clash of ideologies and evolving economic paradigms between the East and the West, asking: Which vision for the world could take the lead in the coming decades? Chapter 5 brings the book to a close by putting the new global economy of the 21st century under the microscope and by examining the significance of economic inclusion as a unifying factor for possibly bringing the visions and ideologies of the East and the West closer together. This is not only to assist countries in achieving the SDGs, but cast a shared vision for the global economy in the 4IR that can advance everyone's quality of life, sustainably.

As you embark on this journey through the pages of *West to East: A New Global Economy in the Making*?, I invite you to critically engage with the arguments presented, reflect on the implications of these shifts, and consider how you, in your capacity, can contribute to shaping a more inclusive and balanced global economic future.

Bloemfontein, South Africa Arno J. van Niekerk

Competing Interests The author has no competing interests to declare that are relevant to the content of this manuscript.

Contents

About the Author

Dr Arno J. van Niekerk is an economist and senior lecturer at the University of the Free State. He reviewed academic journals, written an award-winning book and presented numerous research papers at national and international conferences. He also lectures annually at the Fachhochschule (FHS) University of Applied Sciences in Salzburg, Austria, as part of their masters' programme.

Rise of the East: Economic Implications of the Geopolitical Shift

1

Abstract

This chapter explores the economic consequences of the perceived geopolitical shift from Western dominance towards a more multipolar world, focusing on the rise of the East. It examines the changing landscape of global economic governance, trade, and investment flows, and the increasing importance of technology in shaping international relations. The research reveals a transition from a US-dominated unipolar system to a more balanced, multipolar governance model. This evolution is characterized by the increasing assertiveness of Eastern economies in global institutions and their growing impact on trade, investment, and development patterns. These changes are reshaping the established norms and structures of global commerce and governance. The significance of this research lies in its analysis of the evolving global economic order and its potential impact on international relations. It underscores the need for global governance institutions to adapt to these changes to maintain stability and promote inclusive and sustainable development.

Keywords

Global power shift • Geoeconomics • Geotechnology • Global economic governance

1 Introduction

Progressives in the West, especially after the Cold War, believed that the triumph of liberal values would secure liberal capitalism's role as the sole viable economic model. These free-market ideologues thought it had irreversibly shaped the world. They attributed it to the profound systemic shifts brought about by economic, social, and technological globalization, as well as increasing global interconnectedness. These developments were accompanied by a substantial expansion of

A. J. van Niekerk, *West to East: A New Global Economy in the Making?*,
https://doi.org/10.1007/978-3-031-93267-0_1

international institutions and economic governance mechanisms (Hurrell, 2024). Globalization was seen as giving rise to new and extensive forms of economic integration and governance. The prevailing narrative suggested an inexorable march towards the global dominance of Western capitalist modernity. It proved to be quite illusory. Underlying assumptions prioritized economic prosperity and efficiency, which overemphasized a seeming 'predictable evolution' of institutions and the legitimacy of power derived from ingrained liberal economic and political values.

At the centre of this worldview, the United States leveraged both hard and soft power to position itself as the indispensable nation leading the global liberal order. Never expecting a challenge to their power, the shock and wake-up call to the West came in the form of the 2007–09 global financial crisis (GFC). Since then, traditional perceptions about global order have been challenged in new ways due to the resurgence of geopolitical tensions, the inherent instabilities and inequalities within liberal capitalism, the impact of new and disruptive socialist alternatives (e.g., China's success), social mobilisation (e.g. social media), and ground-breaking technologies. The global community witnessed a remarkable shift in global affairs compared to only a few decades ago. Yet, despite globalization being in retreat, the global impact of changes continued to shape a new global reality. The potential decline of the dollar's hegemony, the axis of economic power moving East, and the rise of Artificial Intelligence (AI) and the Internet of Things (IoT) are all part of this new scenario.

These developments, beginning to surface more distinctly since 2020, may abruptly have changed the world forever, yet it is their economic implications that are crucial in determining whether they are solving (or aggravating) urgent global challenges, such as economic inequality, environmental sustainability, trade conflicts, and geopolitical tensions. Ideally, a shift in the global balance of power ought to correspond with an improved capacity to deal with factors that are making the global economy unsustainable. The reality is that the global liberal order, comprised of embedded economic liberalism and an outward-looking and engaged foreign policy, are being challenged in the twenty-first century by a more plural, diverse and multilateral global order that is not West-centric. Further complicating matters, is the mixture of: (1) moving from dealing with global challenges to dealing with planetary imperatives (i.e., urgent existential threats); (2) geopolitical rivalry that is leading to the politicization and securitization of many aspects of the global economy (i.e., weaponized interdependence); and (3) the challenges of illiberal authoritarianism externally and of nationalist and backlash politics domestically (i.e., internal and external polarization).

According to Hurrell (2024: 221), "new cleavages have opened up between those in favour of continued global integration and global governance on the one hand, and those who reject the opening of borders, the transfer of political authority beyond the nation state, and the promotion of proclaimed universal values on the other". A significant influence in this development has been the expansion of BRICS (Brazil, Russia, India, China and South Africa). Not only have the BRICS countries grown stronger since the organization's founding in 2009, they have also intentionally built partnerships with countries outside of the West's power axis.

While economic integration has increased within the global economy, the rise of different power blocs are fragmenting and realigning economic ties. Competing interests and rival economic agendas have brought the global economy to turbulent waters, yet not without new economic opportunities and developments.

China and India's economic growth, specifically, as well as the rise of the Asian Tigers (Southeast Asian countries) and Japan, has significantly shifted the global economic centre of gravity towards Asia. Their increasing economic integration and trade influence have become particularly observable since the GFC. New opportunities they have brought to the global economic landscape include innovation leadership, expanding markets, diversified trade, new sustainability solutions, cultural exchange, new investment opportunities, as well as new forms of diplomacy and economic cooperation (Attiná & Feng, 2023). Among the countries in Asia, China does stand out as already a global economic giant, albeit in its infancy in terms of development-potential, and is noticeable for its modern version of a largely socialist, state-centric economy, which is unignorably a counter system to Western capitalism. With Russia a strong ally along with key Arab states in the Middle East, China's socialist influence is growing day by day. This is reshaping international alliances.

Before asking whether China's dominance is good or bad for the world, we need to critically evaluate this reality—or new reality—that is taking shape. Indications are that the West is mostly focused on survival, while the East is blossoming and casting a new vision for the world. The implications are weighty, given the opposing forces of market-led capitalism and democracy represented by the West and centralised planning and socialism represented by some leading countries in the East. These implications require in-depth investigation. In addition, an intriguing question is whether this could eventually lead to a kind of synthesis, resulting in an inconceivable blend of capitalism and socialism or some new type of symbiosis. While the future is uncertain, a new reality is most certainly emerging. Seldom have we seen such strong opposing forces on a global scale, signifying, perhaps, a transition from an 'old order' to a 'new order' with significant geopolitical risks. This certainly warrants examination, and it comprises the focus of this chapter.

To give further context, Fig. 1 shows how trade flows and foreign direct investment (FDI) are directly affected by economic crises and geopolitical risks (e.g., after 9/11). After the GFC, global FDI decreased from 3% of global GDP in 2007 to 2% in 2009, while trade also decreased (IMF, 2023). The COVID-19 crisis led to a rapid decrease in global trade and FDI in 2020 and an escalation of geopolitical risk. Although global trade and FDI recovered somewhat, geopolitical risk continued to rise, aggravated by the Ukraine war and the war in Israel and the Middle East.

Figure 1 illustrates that over the past two decades, there has been a slowdown of globalization (also called, 'slowbalization'), which can be linked directly to severe global crises and geopolitical fragmentation. Growing geopolitical tensions and the unequal distribution of globalization's benefits have fuelled rising scepticism toward multilateralism and heightened the attraction of inward-focused policies (IMF, 2024). Brexit, trade disputes between the US and China (especially since

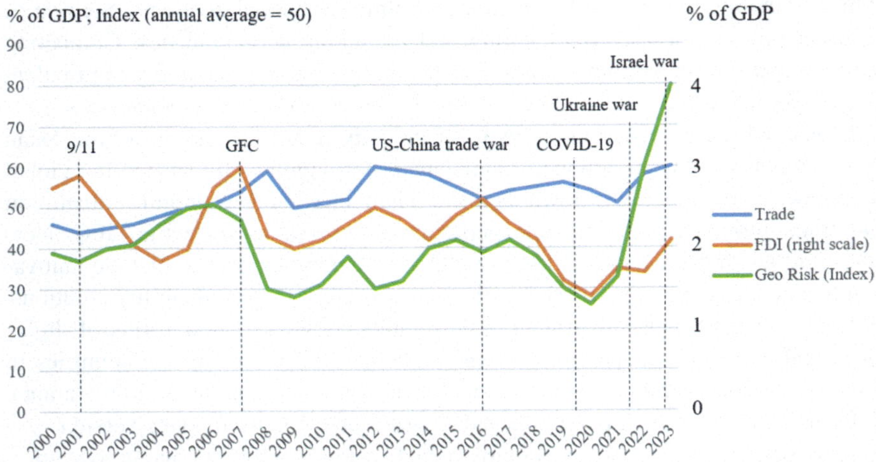

Fig. 1 Global geopolitical risk, trade and FDI. *Source* IMF (2023, 2024), Hassan et al. (2019)

the first Trump administration), Russia's invasion of Ukraine, and the Middle East crisis after 7 October 2023 challenge international relations and risk, triggering a policy-driven reversal of global economic integration, known as geoeconomic fragmentation. This fragmentation spans various channels, including trade, capital, and migration flows. In China, just like in Western countries, government directives are focusing increasingly on replacing imported technology with domestic alternatives to reduce reliance on geopolitical competitors (Mahoney, 2023). Multiple factors have therefore driven this period of slowbalization, yet the fragmentation of trade and capital flows along geopolitical fault lines and the emergence of regional geopolitical blocs introduce new dynamics that could have substantial negative impacts on the global economy.

Now that Donald Trump has again taken Presidential office in the US, the levels of geopolitical and geoeconomic tensions are expected to rise for the remainder of the decade. Given an increased inward policy focus, slowbalization is likely to continue in a climate of heightened global risk. Even if Trump puts an end to the wars—as he promised during his campaign—it would actually add to the tension between the power blocs in the West and the East. This gives further reason to investigate the potential shift in economic and political power from West to East. The momentum has shifted toward the East, but if Trump were to counter some of that shift, it would set the stage for an intriguing decade ahead. Let's examine key economic factors that are playing a telling role.

2 A Shift in Global Economic Power

The staggering growth of the Chinese and Indian economies, in particular, have resulted in a gradual rebalancing of global power since the year 2000. Japan became a leading economy after the 1960s, but now it is no longer so by itself in the East. The economic success of the Southeast Asian countries (especially Singapore, Indonesia, Malaysia, and the Philippines), the Russian resurgence, and the expansion of economic interests by Arab states, such as the United Arab Emirates (UAE) and the Kingdom of Saudi Arabia, coincided with China and India's expansion.

A radically new balance of power is taking shape as particularly the Sino-Russian-Arab strategic partnership embodies the rise of a new 'global multipolar order' that is actively working towards offsetting American influence in the East and Middle East. Almost overnight, reality changed, as pointed out by Chinese President Xi Jinping when he stood at the door of the Kremlin and said, "Right now there are changes—the likes of which we haven't seen for 100 years—and we are the ones driving these changes together" (Aljazeera, 2023). In view of the BRICS+ countries having decided (in Johannesburg) in August 2023 to expand its member countries to include Iran, Egypt, Ethiopia, and the UAE (with 45 others having applied for membership), it appears that the Eastern Axis of power is advancing their plans of creating a counter-economic bloc to that of the West.

Only three decades ago, it was completely different. Immediately after the Cold War, the global economy and geopolitical landscape were largely unipolar, dominated by the United States, both militarily and economically. As the sole superpower of the day, the US played a central role in global institutions like the International Monetary Fund (IMF), World Trade Organization (WTO), the World Bank, and the United Nations (UN). Further consolidating economic power in the West, the creation of the European Union (EU) in 1993 and the euro in 1999 reinforced the unipolar world order reality at the time. For over half a century, the US dollar has been the world's dominant reserve currency and since 1999, it arguably found a partner in the euro. However, this entire era of unparalleled Western supremacy may now be on the verge of dramatic change. The rise of China and other emerging powers in Asia and the Middle East is shifting the balance of power toward a more multipolar world order, where global influence is spread across multiple key players instead of being concentrated in only one or two superpowers.

Before we delve into the changing current reality, let's first consider the road that led to this point. After World War II, a rules-based world order started to take shape with the formation of the UN, World Bank (initially called the International Bank for Reconstruction and Development (IBRD)), IMF, and the precursor to the WTO (founded in 1995), the General Agreement on Tariffs and Trade (GATT). It was at the Bretton Woods Conference in July 1944 where 730 delegates from the 44 allied nations gathered in Washington to regulate the new, post-war international monetary and financial order (Bordo, 2017). The Bretton Woods system that came out of this conference established the rules for commercial and financial relations among the 44 member nations. It was the first example of a fully negotiated monetary order designed to regulate monetary relations among independent states. Under this system, countries were required to maintain their currency values within 1% of fixed parity rates against the US dollar, which was convertible into gold at a rate of US$35 per troy ounce of fine gold (or 0.88867 g of fine gold per dollar) for foreign governments and central banks (Eichengreen, 2021). By signing the Bretton Woods agreement, member countries (mostly Western nations) set up a system of rules, institutions, and procedures to govern the international monetary system. What made this system successful was the emphasis on limited international capital mobility, strict financial regulation to help stabilize currencies, and the dominant economic and financial role of the United States and its currency, the dollar.

Through the Marshall Plan, the US provided foreign aid to Europe and Japan, assisting them in making key economic reforms and rebuilding infrastructure and their industrial bases. It functioned for four years and in 1951, the Marshall Plan was largely replaced by the Mutual Security Act. In doing this, the US outlined its objectives of rebuilding war-torn regions, removing trade barriers, modernizing industry, enhancing European prosperity, and preventing the spread of communism (Carew, 1987). However, given that the Bretton Woods system was based on fixed exchange rates, with currencies pegged to the US dollar and convertible into gold, it came to an end in 1971 due to the depletion of US gold reserves, and the declining ability of the US to maintain the dollar's convertibility to gold. The pressures of inflation, trade deficits, and global speculation ultimately led President Nixon to suspend the dollar's link to gold (i.e., the gold standard), marking the official collapse of the Bretton Woods framework and the transition to a floating exchange rate system. This marked the beginning of a more decentralized international economic order, where currency values were determined by market forces, introducing new volatility but also flexibility.

Triggered by the OPEC embargo, the 1973 oil crisis and the 1979 energy crisis caused a steep rise in oil prices, which resulted in stagflation (a combination of stagnation and inflation) in many Western economies (Schumacher, 1985). These crises highlighted the vulnerability of advanced economies to global supply shocks, emphasizing the need for more robust global economic coordination. Developing countries, especially those dependent on oil imports, suffered from high prices and debt, leading to growing disparities between rich and poor nations (Lin, 1984). In the wake of rising global inequality, the liberalization of financial markets gained momentum in the 1980s and 1990s. This led to a significant rise in global capital flows and deregulation in financial markets and financial institutions. Significantly, this era also saw the expansion of the Eurodollar market (dollar-denominated deposits outside the US), which contributed to the US dollar remaining the dominant global reserve currency, despite the absence of a gold standard. In addition, since the 1970s, the US made a strategic agreement with Saudi Arabia—the world's largest oil producer at the time—to price and trade oil exclusively in US dollars (Gökay, 2021). In return, the US promised military protection and support for Saudi Arabia. This was the start of the 'petrodollar' or 'petrodollar recycling' whereby most global oil transactions were conducted in US dollars, creating a significant demand for the currency as countries around the world need to hold large dollar reserves to purchase oil. This helps to support the value of the dollar in global markets.

Triggered by capital flight, currency devaluations and excessive foreign debt, the 1990s and early 2000s witnessed a series of financial crises in emerging markets, including the Asian Financial Crisis (1997), the Russian Debt Crisis (1998), and the Argentine Crisis (2001). The IMF, which is based in the US, played a central role during this period by providing bailout packages to affected countries, albeit with controversial structural adjustment programs attached (Emeagwali, 2011). It was a time when neoliberal policies were promoted, globalization expanded, and the ideology of the Washington Consensus directed countries' economic growth paths. The continuing trend of financial liberalization and deregulation culminated in the global financial crisis of 2007–2009. Excessive risk-taking exposed vulnerabilities in the global financial system, which should have been detected by especially the IMF through its country surveillance initiatives. Central banks then had to implement unconventional monetary policies, such as quantitative easing (i.e., purchasing of large amounts of government bonds or other financial assets to inject money directly into the economy to lower interest rates and encourage borrowing and investment, thus increasing money supply and aiming to stimulate economic activity and prevent deflation).

Importantly, the crisis renewed calls for reforms in the international monetary system and raised questions about the role of the US dollar as the global reserve currency, although the dollar maintained its dominant position. The Eurozone Crisis succeeded the GFC from 2009 until 2014. It involved sovereign debt problems, economic imbalances within the Eurozone, and the inability of some countries to adjust to external shocks due to the common currency. Portugal, Italy, Ireland, Greece and Spain (PIIGS), in particular, have accumulated significant public debt before the GFC and became unable to finance their debts (Beker, 2014). The crisis was especially severe because, as part of the Eurozone, they could not devalue their own currencies to make their exports more competitive or control their monetary policy, as they were tied to the euro. They relied on bailout packages from other Eurozone countries, placing national budgets under tremendous pressure. The recession Europe experienced stressed the challenges of economic integration without fiscal union and led to drastic reforms in European governance. While the crisis was eventually contained, it exposed weaknesses in the Eurozone's structure and had lasting social and economic effects.

The post-crisis era has been characterized by ultra-low interest rates and, especially after 2020, the proliferation of thousands of cryptocurrencies. The latter occurred as part of the formation of decentralized networks that typically make use of blockchain technology, and are not controlled by any central authority like a government or central bank (Skaf, 2023). Digital/virtual currencies like Bitcoin (started in 2009) and Ethereum (2015) became well-known examples. They are used for online transactions, investments, and as a store of value. A further development around the future of money and the international monetary system, is the emergence of Central Bank Digital Currencies (CBDCs). These are digital versions of a country's official currency, issued and regulated by the central bank. Unlike cryptocurrencies, CBDCs are centralized and represent a direct liability of the central bank, making them a secure and official form of digital money (Pandey, 2022). They are designed to facilitate secure, efficient, and cash-like transactions in the digital space, and can be used by individuals, businesses, and financial institutions.

Apart from the digital currency revolution during and after the disturbing COVID-19 pandemic (2020–2022), the post-GFC era was marked by the rise of China and other emerging markets, and China's push for the internationalization of the yuan, which has led to greater multipolarity in the global monetary order (Geeraerts, 2011). New financial institutions, like the Asian Infrastructure Investment Bank (AIIB) and the BRICS New Development Bank, started to reflect efforts by emerging economies to create alternatives to Western-led financial institutions. These represent the first steps in developing an alternative (or moving away) from the rules-based international order[1] that originated at Bretton Woods. With global economic governance institutions like the IMF, World Bank and even the WTO being criticized for becoming outdated, a new multipolar system started to emerge in conjunction with decentralized financial systems using cutting-edge technology.

Today, a new global reality is evident. The international monetary order has evolved significantly since the end of the Bretton Woods system. The shift from fixed exchange rates to floating rates (1970s), the rise of capital mobility (1980s), the emergence of new economic powers, and the increasing role of digital currencies (2000s) have all contributed to a more complex and dynamic global financial landscape. Despite the US dollar's hegemony, the international system has become more decentralized and multipolar, reflecting the changing balance of global economic power.

This multipolarity is multilayered and culturally diversified. China, in particular, is no longer the developing nation it once was and is growing more assertive by the day. With Beijing leading arguably the world's most successful economy, its influence on global governance will continue to increase significantly (Song & Ai, 2023). China is gradually assuming the role of a new hegemon. It is already less interested in a radical revision or reform of the current world order, and more interested in replacing it. Partnering with its other BRICS+ members, strategic decisions were made at the 2024 BRICS Summit in Kazan, Russia, to adopt an independent international payment system (BRICS Pay as an alternative to SWIFT) based on their national currencies to steer away from the use of dollars (Lee, 2024). The BRICS+ member nations, representing roughly 35% of the global economy and almost 50% of the global population, issued the Kazan Declaration, which calls for the "elimination" of "unilateral economic sanctions and secondary sanctions that are contrary to international law" (Lee, 2024: 1). They expressed

---

[1] The rules-based Bretton Woods system refers to rules and norms for international cooperation set through the UN, IMF, World Bank and WTO, which guide trade agreements, environmental standards, human rights, etc.

concerns over "unilateral coercive" sanctions that have, what they consider, "disruptive effects" on the global economy and agreed to examine the "establishment of an independent cross-border settlement and depository infrastructure" (Ministry of External Affairs, 2024: 1).

One aspect that China often promotes is the idea of multilateralism as a counterbalance to what it sees as unilateral dominance by Western countries, particularly the US. However, China's version of multilateralism tends to focus on state sovereignty and non-interference, which contrasts with Western approaches that emphasize human rights and democratic governance (Attiná & Feng, 2023). These socialist aspirations are a grave concern for the West and set up an era of intense ideological competition at global level. An example of such aspirations is how China advocates for a concept called 'cyber sovereignty,' which accentuates the control of states over their internet infrastructure and digital spaces. This contrasts with the more open and decentralized approach favoured by the West, and it supports China's own heavily regulated digital space. China is striving to lead in technology governance, especially in emerging fields like 5G, AI, quantum computing, and digital currencies. Through initiatives such as 'Made in China 2025' and the Digital Silk Road, China seeks to set global standards in these industries, reducing dependency on Western technology and shaping the digital economy of the future (Hillman, 2021). Having strong partners through BRICS+, its growing influence is reshaping the geopolitical and geoeconomic landscape.

In this respect, another factor causing a shift in global economic power, as pointed out by Nedelcu (2024: 1979), is the fact that "the twenty-first century may witness a transformative shift in strategic thinking from traditional geopolitics and geoeconomics to the new realm of geotechnology". Technological innovations are becoming ever more powerful factors in decision-making processes traditionally dominated by geopolitical or geoeconomic considerations. It is in this context that China and other members of the BRICS+ countries, such as India, Russia and the UAE, have detailed plans to take the lead in the Fourth Industrial Revolution (4IR). The merging of physical, digital and biological areas is placing a growing demand on technology-centred thinking in world affairs. Opening up the way for a more interconnected and interdependent international system, the 4IR's potentially disruptive technology-driven capabilities may actually contribute towards democratizing global governance through its transformative impact on shaping (or reshaping) the future of international relations. All these factors will reorient global interactions as their impact on the shift in global economic power grows. These changes suggest an increasingly decentralized and competitive global order, with continuing challenges to the traditional, Western-led system.

3 A Shift in Global Trade and Investment Flows in Key Asian Economies

Asia is emerging as a central hub for attracting and stimulating international trade and investment flows, not just from an intra-regional perspective, but also inter-regionally. The twin priorities of developing regional supply chains and reducing reliance on Western markets, have resulted in the Regional Comprehensive Economic Partnership (RCEP) (Australian Government, 2022). Being the world's largest trade agreement, this covers key Asia–Pacific economies and has stimulated economic growth by lowering tariffs and streamlining trade processes. China, in particular, is constantly pushing for trade rules that benefit its economic model and that is aligned with its state-directed economy. China sees the value of the RCEP as being a counterbalance to US-led trade agreements like the Trans-Pacific Partnership (TPP). As such, South-South trade flows have become a significant part of global commerce as Asia is expanding its trade ties with Latin America, Africa, and the Middle East. This is creating new growth avenues for Asian economies.

One relatively new example of this is the increased trade in energy. Asian countries, especially China and India, are investing heavily in energy projects across Africa and the Middle East (He & Guo, 2019). This allows Asia to secure its energy needs, while producing countries diversify away from Western markets. Even in the area of digital trade growth, companies like Alibaba, Tencent, and Flipkart are investing in digital infrastructure and logistics across Asia, which is enhancing regional trade efficiency (Heeks et al., 2024). This is contributing significantly to making Asia the world's largest e-commerce market, driven largely by China, Japan, South Korea, and Southeast Asian nations. China alone represents nearly half of global e-commerce sales, with total online retail sales amounted up to almost US$2.2 trillion in 2023, and the US ranking second with sales less than half of that: US$981 billion (Buchholz, 2024). Given Asia's high internet penetration and smartphone usage, rapidly growing middle class (increased demand), innovative e-commerce models like live-stream shopping and super apps, it has become the global leader in e-commerce.

Asian countries, especially China, are increasingly investing in emerging markets through projects like the Belt and Road Initiative (BRI). Having been launched in 2013 by President Jinping, the BRI is a global infrastructure and economic development project that plans to invest in more than 150 countries and international organizations (World Bank, 2018). By establishing land and sea networks, the project aims to enhance connectivity and trade across Asia, Europe, Africa, and even Oceania and Latin America by developing transportation, energy, and digital infrastructure to enhance regional integration, trade, and economic growth. It has mainly two components:

- The Silk Road Economic Belt: A network of overland routes that connects China to Europe through Central Asia. It consists of railways, highways, and pipelines, intended to streamline the flow of goods and energy.

- The 21st Century Maritime Silk Road: A maritime route that links China to Southeast Asia, South Asia, the Middle East, and Africa through a series of ports and coastal infrastructure, which facilitates trade and reduces shipping costs.

The BRI has resulted in numerous completed projects, including high-speed rail links in Southeast Asia, major ports in Pakistan (Gwadar Port) and Sri Lanka (Hambantota Port), and energy plants in Africa (e.g., the Kariba South Bank hydropower expansion in Zimbabwe, the Lamu coal power plant in Kenya, and the De Aar wind power project in South Africa). Chinese investments and contractual cooperation for 2023 in the 149 countries currently part of the BRI amounted to about 212 deals worth US$92.4 billion (Wang, 2024). This is 18% more compared to that of 2022 (US$74.5 billion). Investment growth was particularly evident in technology (by 1046%) and in metals and mining (158%). Africa also became the largest recipient of Chinese engagement, overtaking Middle Eastern countries. In terms of trade growth, the BRI increased bilateral trade between China and the BRI countries since its inception by between 6.37 and 8.53% more than that with the non-BRI countries (De Marcos, 2022). Banach and Gunter (2023) highlight an important point in that, as the level of China's control at different ports in BRI countries increase, total trade also increases with China.

The global network of Chinese owned, operated or built port terminals grew from 10 in 2013 to 75 in 2023, giving China extensive influence at ports around the world, which is reorienting trade routes (Budanov et al., 2024). For example, Chinese firms are buying an ever-greater share of their goods from countries where China has control over terminal operations. Banach and Gunter (2023) found that China's port network is playing a significant role in reshaping international competition because of the observable changes in exports, imports, and total trade flows after signing a terminal operating contract. For example, where Chinese firms manage ports, they seem to shift the host countries' trade flows toward China and away from previous trading partners. In reality, China's export dominance and supply chain integration have reoriented global production networks, reducing the centrality of Western economies in global trade (Schindler et al., 2024). Moreover, Asian markets are now major export destinations for developing countries, replacing the West as the primary driver of trade growth. The downside of this, however, is that it might cause China and its Asian counterparts to lose market access in the West. In a study by Qiu et al. (2024), the authors found that between 2017 and 2023 quarter-on-quarter trade volumes grew around 2.5% more slowly for geopolitically distant countries relative to those that were closer. This confirms that geopolitical alignment has a significant effect on trade volumes, which can be expected to increase as tensions between East and West grow.

According to Wang (2024), potential future engagements of the BRI is expected in mainly six project types: manufacturing in new technologies (e.g., batteries), renewable energy, trade-enabling infrastructure (including pipelines and roads), ICT (e.g., data centres), resource-backed agreements (e.g., mining, oil, gas), and high visibility or strategic projects (e.g., railway). The BRI is primarily funded by

Chinese banks and development institutions, such as the China Development Bank, the Export–Import Bank of China, the Silk Road Fund, as well as AIIB funding (Hu et al, 2021). The initiative operates through partnerships with local governments, public–private investments, and contracts awarded to Chinese companies. A core part of the BRI is China's Digital Silk Road that advances its digital infrastructure (e.g., fibre optics, 5G networks, data centres, and e-commerce platforms) across Africa, Southeast Asia, and the Middle East. Through digital connectivity, China is expanding its influence in global digital markets, setting new standards for innovation, and fostering economic partnerships beyond physical trade routes. The BRI is the centrepiece of China's foreign policy and displays its unique economic development model: attaining long-term economic growth through systematic and broad-based investments in different types of infrastructure assets (Javed et al., 2021). This contrasts with the neoliberal approach advocated by the West and its Bretton Woods institutions, which emphasizes short-term export-driven and domestic consumption development models.

India is also playing a significant role in the shift of global trade and investment flows, reshaping regional and international economic dynamics. India has made remarkable progress over the past two decades with regards to regional trade alliances, strengthening its trade relations with countries across Asia, Africa, and the Middle East. One focal point is its partnerships within the South Asian Association for Regional Cooperation (SAARC), which underscores India's intention to integrate more deeply with Asian economies (Shaheen, 2013). India keeps on advancing bilateral agreements with countries in Southeast Asia, Japan, and South Korea, and trade agreements with the EU and the United Kingdom (UK). Recent agreements with the UAE and Australia further aim to increase its exports and enhance supply chain links within the region. This is part of India's 'Look East' policy that started in the 1990s, which is altering regional power dynamics (Naidu, 2006). India is strengthening its economic and security ties with East Asia in response to China's growing influence in the region. Under Prime Minister Narendra Modi in 2014, the 'Look East' policy evolved into the 'Act East' policy, emphasizing proactive, action-oriented engagement to deepen economic, political, and security connections (Bajpaee, 2023). This policy broadened India's scope to include the wider Asia–Pacific, which includes Japan, South Korea, Australia, and even the US through platforms like the Quad (Quadrilateral Security Dialogue).

In 2022, India ranked as the world's fifth largest economy by GDP (current US$), 15th in total exports, and 8th in total imports. India's travel exports (year-on-year growth in travel during the first quarter of 2023) increased by 206% while commercial services increased by 23%—well ahead of the US, UK, China, Germany, and France (WTO, 2023). India has strengthened collaborations in digital trade, cybersecurity, and FinTech with especially Asian and Middle Eastern countries, contributing to its growing digital economy and supporting new investment flows. Having become a global leader in IT with a rapidly growing market for digital services, India has attracted significant investment in technology, which contributed to its influence on digital trade and innovation (UNCTAD, 2024).

The International North–South Transport Corridor (INSTC) is a significant trade route for India, linking it with Russia via Iran and Central Asia. It connects the Indian Ocean and Persian Gulf with the Caspian Sea via Iran, and then onwards to northern Europe via St. Petersburg in Russia. This multi-mode network of ship, rail, and road route for moving freight facilitates faster, cost-effective trade with Europe and Eurasia (Sahakyan, 2020). The INSTC aligns India with growing Eastern markets and enhances its logistical capabilities in the region. India stands, ironically, as perhaps the most Western-aligned BRICS+ member, among the greatest beneficiaries of the group's evolution and expansion. India has established strong connections with many new BRICS+ members: Egypt is an expanding trade and security partner in the Middle East, the UAE is one of India's most significant partners overall, and its relationship with Ethiopia is one of the longest and closest in Africa (Kugelman, 2024). The original BRICS members also offer key advantages for India. Through BRICS, Delhi can affirm its enduring support for Russia, despite Western efforts to isolate it. Furthermore, by collaborating with rival China within BRICS, it allows India to cautiously ease tensions with Beijing, reinforced by a recent border patrolling agreement announced just before the 2024 summit. BRICS also provides a platform for India to pursue its core foreign policy of strategic autonomy, allowing it to balance relationships with a wide range of geopolitical actors without formal alliances. Thus, India plays a vital strategic role as it aims to balance its ties with the Western and non-Western worlds.

The INSTC also benefits Russia significantly, as it enhances its trade connectivity, economic ties with key regions, and its strategic positioning in global logistics. For instance, the INSTC offers Russia a direct trade route to India, Iran, Central Asia, and the Middle East, providing a better alternative to traditional routes via the Suez Canal. This allows Russia to diversify its trade partners, particularly as it seeks alternatives to Western markets due to sanctions and geopolitical tensions. The INSTC is vital for the required reconfiguration of supply chains in Eurasia. According to Vinokurov et al. (2022), given the drastic increase in Russian exporters, importers and freight-forwarding companies' needs in alternative logistical opportunities, it has become invaluable to the Russian economy. The INSTC is promoting Eurasian intra- and transcontinental connectivity, reducing export costs, developing new production niches, and is starting to realize the Caspian region's transit potential. This has contributed to the IMF ranking Russia, despite its war with Ukraine, as the fourth largest economy in the world based on purchasing power parity (PPP) (IMF, 2024). In October 2024, Russia's GDP amounted to 3.55% of global GDP in PPP terms. It outperformed Japan (3.38%) and overtook the UK, France, and Germany. With China (18.8%), the US (15%) and India (7.9%) ahead of Russia, the world's leading economies included three BRICS countries at this point in time.

The geopolitical realignment from West to East is reshaping global investment flows. Companies are actively diversifying their sourcing and production strategies to reduce dependence on certain suppliers and mitigate geopolitical risks (Rosenberg & Sandrini, 2023). This has led to a shift towards more regional markets and a reassessment of cost structures as supply chains become shorter. In terms of green investments, Asia is becoming a global leader in renewable energy investment, particularly in solar and wind technologies. In Southeast Asia alone, private sector investments to stimulate the green economy transition increased by 20% to US$6.3 billion in 2023 compared to 2022 (US$5.2 billion) (Hardcastle et al., 2024). Countries like China, Japan, and India have rapidly scaled up renewable capacity, attracting investment flows in sustainable energy projects. China invested $890 billion in clean energy sectors in 2023 (similar to the GDP of Switzerland) and issued $80 billion worth of green bonds in the same year, making it the world's largest issuer (Zhang et al., 2024). In Japan, the number of greenfield investments in 2022 increased by 11.4% from the previous year (after four years of stagnation), and kept increasing in 2023 (JETRO, 2023).

In India, the value of sustainable investments in 2021 was US$19 billion and grew to about US$70 billion by 2023 (Joshipura et al., 2024). In the same year, India's green bond issuances were worth around US$21 billion—of which 84% was from the private sector. Except for China, this is more than any other emerging Asian market. In addition, India is one of the fastest-growing renewable energy markets in the world. Asia has become the primary growth region for global energy consumption, driving demand for both traditional and renewable energy sources. This has made Middle Eastern oil producers and renewable energy markets more focused on Asia (Mirza et al., 2023). China, in particular, is positioning itself as a global leader in renewable energy production, exporting solar and wind technologies. This shift is reshaping global energy markets and challenging Western dominance in clean technology innovation.

A targeted emphasis in Asia is decarbonizing supply chains. As sustainability grows in importance, Asia's supply chains are adapting to greener practices, including circular economy initiatives. This trend appeals to environmentally conscious investors, especially from Europe, who prioritize low-carbon suppliers. According to a report by McKinsey & Company (2022), investors in especially manufacturing investment do this by including carbon prices into all their investment decision-making. They do it as an innovative test to see whether the investment decision would be different given a certain carbon price, or they build this into a certain decision as a core assumption. While net-zero transition and decarbonization have become an integral part of every country's economic growth and development strategy, the ASEAN (Association of Southeast Asian Nations) region faces the fundamental long-term challenge of meeting rapidly rising demand safely and cost-effectively. Yet, data shows that in aggregate, a 32% rise in renewable investments resulted in a 36% increase in ASEAN's renewable output, which is most prominent in Singapore (150%), Indonesia (140%), the Lao People's Democratic Republic (127%), and Malaysia (101%) (ADB, 2023). This is coupled with an increase in employment as the focus moves away from energy-intensive sectors and more towards an increase in output and exports of non-energy-intensive commodities. Through the use of greener and smarter technologies—such as AI, blockchain and the Internet of Things (IoT)) optimize energy use, reduce waste, and enhance efficiency—as well as using electric vehicles (EVs) and more rail transport, Asian economies are decarbonizing their supply chains and transitioning supply chain logistics.

Policy-makers across Asia are actively working on reducing trade barriers in the renewables supply chain to make the green transition easier (Zhang et al, 2022). Eliminating trade barriers, especially within the renewables global value chain, would significantly lower costs and make it easier to achieve net zero. To support this, Asian countries are engaging in regional and global partnerships, such as the Asian Development Bank's (ADB) support for low-carbon infrastructure projects, and Japan's collaboration with ASEAN on green technology adoption. Such investments aims to counterbalance current global value chains' fragmentation and decoupling. However, what makes it complex is that, particularly after the COVID-19 pandemic, many companies are shifting production facilities closer to or within Asia, reducing reliance on Western markets due to US-China tensions (Joshipura et al., 2024). Regional supply chains allow for greater stability and diversification. Hence, Asia is now focusing more on supply chain resilience (through diversification) and localized production, especially in critical sectors like semiconductors and pharmaceuticals. This shift reduces exposure to external disruptions and enhances regional control.

4 Technological Transfer and New Innovation Hubs

The rivalry between specifically the US and China is heavily focused on technological development. This competition over technology is affecting trade, investment, and the control of data governance and cybersecurity (Hurrell, 2024). Both countries vie for dominance in critical technology areas like 5G, AI, and semiconductor manufacturing. This rivalry is reshaping the global economic landscape, creating new policies, alliances, and concerns for countries worldwide. Translating the trade war into a 'tech war', the US has imposed export restrictions on advanced technology and sensitive materials to prevent China from gaining access to cutting-edge technology—even to the extent of restricting Chinese companies like Huawei and SMIC to having access to US-origin technology (Mirrlees, 2024). The 4IR-competition is driving both nations to 'decouple' from each other in critical technology supply chains. The US has encouraged firms to relocate manufacturing out of China. Both countries have intensified cybersecurity measures, each accusing the other of cyber espionage and intellectual property theft. The US views Chinese tech companies as potential national security risks, banning platforms like TikTok and WeChat. China has similarly blocked US tech firms such as Facebook and Google from operating in the country.

With particularly China and India, as tech giants, investing heavily in AI, 5G, and digital currencies, Asia is fast becoming a leader in technological innovation. This shift has led to increased competition, not just with the US but also with Europe, over technology standards and data sovereignty (Mahoney, 2023). It is fostering new alliances and is driving a fragmented global digital landscape, especially as technology cooperation is expanding between Asian countries and emerging markets, creating innovation hubs in a variety of technology types. The following are among the leading tech hubs in the East:

- *Shenzhen (China)*: known as 'China's Silicon Valley', it is a hub for hardware and consumer electronics. It houses tech giants like Huawei, Tencent, and DJI and has a thriving ecosystem of start-ups and manufacturers focused on IoT, AI, and robotics. Alongside Shenzhen, Beijing is China's AI, cloud computing, and big data hub, hosting companies like Baidu and Lenovo and numerous AI research facilities. Shanghai, a third Chinese city, focuses on FinTech, e-commerce, and blockchain technology, with large players like Ant Group and the Shanghai Data Exchange leading digital innovation in finance and logistics.
- *Bangalore (India)*: known as 'India's IT capital' and for some, 'The Silicon Valley of the East', it is a global hub for software development, AI, and biotech. Major tech companies like Infosys, Wipro, and Google have substantial research and development (R&D) centres here, as well as many AI and machine learning start-ups. Microsoft and Amazon choose Bangalore for their Indian headquarters. The city's vibrant cultural foundation, with a strong emphasis on education and innovation, has consistently fostered tech talent. Its many engineering colleges yield a continual flow of skilled professionals, making talent acquisition relatively easy (Jones, 2023).
- *Bangkok (Thailand)*: is advancing in biotech, health tech (healthcare innovation), IoT, and digital innovation as part of the 'Digital Thailand' initiative. Bangkok and Chiang Mai are attracting a steady stream of foreign investment to Thailand given infrastructural developments, easier business registration and attractive tax incentives for tech enterprises.
- *Jakarta (Indonesia)*: is becoming a key hub for e-commerce and digital finance in Southeast Asia, driven by a rapidly growing middle class and digital economy. Major regional start-ups like Gojek and Tokopedia (now GoTo) are based in the city. This digital wave is the result of the Indonesian government having introduced the '1000 Start-ups Movement' aimed at fostering a robust start-up ecosystem. The initiative facilitated mentoring, funding and scaling for emerging tech businesses, making Jakarta home to numerous 'unicorns' (start-ups valued at over US$1 billion).
- *Kuala Lumpur (Malaysia)*: has given birth to several prominent tech companies, such as Grab, a super-app that unseated Uber within the region, and to tech parks like the MSC Malaysia, specifically designed to nurture ICT companies. The city's vibrant cultural diversity and strong sense of interconnectedness have cultivated a dynamic exchange of ideas and innovations (Dabla-Norris et al., 2023). This stemmed from the government's 'Malaysia Digital Economy Blueprint', which outlined an ambitious roadmap for Malaysia to become a technology-driven nation. Another city, often called Malaysia's 'Silicon Valley' is Cyberjaya, which is a smart city focused on AI, cybersecurity, and IoT, with a strong presence of global IT companies and a burgeoning start-up ecosystem.
- *Seoul (South Korea)*: is a leader in ICT, mobile technologies, and semiconductors. It hosts global tech giants like Samsung and LG and has a vast array

of initiatives supporting AI, autonomous vehicles, and 5G technologies. The government's 'Digital New Deal,' a multi-billion-dollar initiative, has been pivotal in strengthening digital infrastructure, advancing AI and big data research, and fostering start-ups in Seoul. One of the city's bustling districts, Gangnam, is home to a myriad of start-ups, tech incubators, and innovation hubs. Rivalling Seoul as South Korea's 'Silicon Valley,' is Daejeon, which specializes in research-based technologies that include robotics, materials science, and biotech. It is home to the Daedeok Innopolis, a cluster of R&D centres and research institutions.
- *Tokyo (Japan)*: is Japan's hub for robotics, AI, and FinTech. It is home to tech giants like Sony, SoftBank, Nintendo, and Hitachi, as well as R&D in advanced robotics and IoT technologies. As part of the government's 'Society 5.0' initiative, it seeks to move beyond traditional frameworks, paving the way for an era where digital transformation is closely aligned with societal well-being (Jones, 2023). The city's tech districts/hubs, such as Akihabara, display its vibrant electronics and innovation culture, teeming with inventors, entrepreneurs, and tech devotees.

These innovation hubs are just one indication of developing economies in the East, like China and India, that are experiencing rapid economic growth and technological advancement, which is presenting a challenge to the traditional dominance of Western nations. This shift in the global landscape, brought about by a sort of 'technological transfer', is driven by several factors—each playing a distinct role in reshaping innovation, economic capabilities, and strategic influence. Firstly, there is a commercial and industrial technology transfer. Asian economies, including China and India, have increasingly acquired advanced manufacturing, energy, telecommunications, and transportation technologies from Western firms, either through FDI, joint ventures, or acquisitions. This has enabled them to become competitive in sectors like automotive manufacturing, electronics, and aerospace. For example, Jaguar Land Rover has been acquired by India's Tata Motors in 2008. This acquisition provided Tata Motors with access to advanced technologies and engineering expertise, contributing to its growth and competitiveness in the global automotive market (Smith, 2013). Overall, trends like these have enabled Asian countries to improve productivity, infrastructure, and industrial capabilities at a grand scale.

Secondly, there has been a digital and information technology transfer. This includes the transfer of software, digital infrastructure, and data management technologies, involving cloud computing, cybersecurity, AI, and telecommunications. Through partnerships with Western firms like Nokia, Ericsson, and Qualcomm, Asian companies have gained access to AI, big data analytics, and 5G infrastructure. China, in particular, has advanced its telecom sector by incorporating Western technology, resulting in companies like Huawei and ZTE Corporation becoming global 5G infrastructure leaders (Erie & Streinz, 2021). They and many other large Chinese tech companies have developed partnerships and intellectual property exchanges to enhance their AI capabilities.

In the same way, South Korean companies like Samsung and LG, and Japanese companies like NTT Docomo and Rakuten Mobile have become leaders in 5G technology (Kim et al., 2023). Apple, for instance, collaborated with Asian suppliers in China and South Korea, which led to a transfer of skills in high-volume production, component manufacturing, and supply chain management. This resulted in rapid advancements in local smartphone brands like Xiaomi, Oppo, and Vivo. There has also been academic partnerships and research exchanges between Western institutions (such as MIT and Stanford) and Asian universities (like Tsinghua and NUS), which contributed significantly to knowledge transfer in AI and machine learning (Gaofeng et al., 2021). This sharing has accelerated AI research and development in Asia.

Thirdly, there has been a green and clean technology transfer. This includes renewable energy technologies, carbon capture, waste management, and other sustainable innovations that support global climate goals and enhances energy security as economies strive towards decarbonization. China's BRI includes investments in green energy projects that involve technology partnerships with Western clean energy companies such as Siemens (Germany), GE Renewable Energy (US), and Vestas (Denmark) (Lewis, 2023). India, as well, has formed collaborations with TotalEnergies in France and BP in the UK, among others, to increase solar energy capacity through the transfer of photovoltaic technology and battery storage solutions (Hayashi, 2020). Initially, the US and Germany were leaders in solar technology, but partnerships and knowledge sharing enabled China to rapidly scale up its production capacity and become the largest producer of solar panels (80% of global market share), and India not far behind (Smith, 2013). Today, Chinese companies like Jinko Solar and Trina Solar lead global production, followed by Indian companies like Waaree Energies and Rayzon Solar (Walton, 2024; Zhu, 2024).

Fourthly, there has been a technology transfer in urban planning and agricultural innovation. Western smart city innovations, like sensor networks and IoT-based urban solutions, have been adapted in Asian cities (e.g., in Seoul, Mumbai, and Ho Chi Minh City, Vietnam). Singapore, for example, worked with companies like IBM on its 'Smart Nation' initiative, while cities in China like Shanghai and Beijing have partnered with Western firms like IBM, Cisco, and Ericsson for smart city technologies, making them leaders in urban IoT (Frana, 2024). In addition, Asian companies, mostly in South Korea and Japan, have adopted Google and Amazon's smart home technology and further innovated in smart home products, integrating them into their own product ecosystems. The agricultural technology transfer involves the exchange of advanced agricultural methods, equipment, and biotechnology to improve productivity and food security (Lee et al., 2021). Precision farming technology, such as drones and IoT-based soil monitoring, is being introduced in parts of Asia, including India and Southeast Asia, through partnerships with Western agricultural firms like John Deere (in India and China), Bayer (in India, Indonesia and Philippines), and Trimble (in India, Japan and South Korea), among many others.

Lastly, there has been a financial technology (FinTech) and e-commerce transfer. This applies to the transfer of technologies related to digital payments, banking, and financial services in how it has accelerated digital financial inclusion and modernized banking systems. China has integrated FinTech innovations rapidly, while other Asian countries such as India, Singapore, Indonesia, and Vietnam have attracted Western FinTech investments to strengthen digital banking and enhance financial access in emerging markets (see Table 1). Their growing digital economies, supportive regulatory environments, and large, young, tech-savvy, and underserved populations have motivated this (Kshetri, 2020). Partnerships and knowledge sharing in FinTech have been central to Asia's digital payment growth. American companies like PayPal and Visa have influenced payment solutions in Asia, contributing to the development of super-apps like WeChat Pay and Alipay in China, and GCash in Southeast Asia (Chodak, 2024).

Attiná and Feng (2023) observe that as developing societies like China and India gain economic power, they may exert pressure to increase technology transfer from advanced societies. This pressure may lead to demands for reduced patent durations or more favourable terms for technology sharing. It appears that emerging economies in the East are strategically using FDI to acquire advanced technologies. Governments in countries like China have introduced policies that promote foreign investment in key technology sectors (IMF, 2023). These policies often include conditions for technology transfer or the establishment of joint ventures with local companies, allowing them to absorb new knowledge and expertise. In view of geopolitical rivalries, especially between the US and China, concerns have increased about the transfer of sensitive technologies, which could reinforce the military capabilities of rival nations or compromise national security. This has led to tighter export controls and restrictions on investments in strategic sectors.

Technology transfers are clearly shifting the global centre of innovation eastward. They allow emerging Asian economies to bridge development gaps and compete globally, especially in high-tech and sustainable industries. Asia's rapid growth in technological advancement is creating new markets for digital economies, which in turn stimulate economic development across the region (Erie & Streinz, 2021). Asian countries are increasingly working together to develop technology standards in areas such as cybersecurity, telecommunications, and AI as part of gaining a competitive advantage. These alliances reduce dependence on Western technology protocols and paves the way for Asia-led tech ecosystems—i.e., networks of technology innovation, entrepreneurship, and investment driven primarily by China, South Korea, Japan, Singapore, and India. What the East and the West share is how central technological innovation is to their visions for the future.

While the Sino-American rivalry (especially the tech war) is expected to intensify, China has already been outpacing the EU since 2017 in terms of percentage of GDP spent on R&D and in terms of world gross expenditure (Stec, 2020). China's strong commitment to advancing its technological capabilities places pressure on Europe to maintain competitiveness in high-tech industries, which could affect global trade dynamics, supply chains, and the interoperability of digital

Table 1 Western FinTech investments in Asian countries

Area of investment	Western investors	Corresponding Asian companies (investment recipients)
Digital payments and mobile wallets	Visa Mastercard PayPal Google Pay	Go-Jek (Indonesia), Pine Labs (India), BillDesk (India) GrabPay (Singapore), Razorpay (India) Pine Labs (India), GoPay (Indonesia) PhonePe (India), BharatPe (India)
Lending, microfinance and neobanking	Goldman Sachs Kiva Revolut N26	KreditBee (India), Toss (South Korea) Ahon sa Hirap, Inc. (Philippines), MBK Ventura (Indonesia) Expanding own operations in Japan, Singapore and India Strategic partnerships across Asia and direct expansion
Blockchain and cryptocurrency	Coinbase Ventures Ripple	CoinSwitch Kuber (India), Bitkub (Thailand) Tranglo (Malaysia), BCRemit, Novatti (both in Philippines)
InsurTech and RegTech	AXA Munich Re Thomson Reuters Refinitiv	Qoala (Indonesia), Coverfox (India) ZhongAn Online P&C (China) KYC (Know Your Customer) and AML (Anti-Money Laundering) compliance tools in Singapore and Hong Kong
Wealth management and Robo-advisory	Schroders BlackRock	Nutmeg and Wealthfront (both in Singapore) StashAway (Singapore)
Remittances	Western Union TransferWise (now Wise)	UnionBank (Philippines), CIMB (Malaysia), Bank of China Direct expansion of remittance services among expatriate populations in Philippines, India, Vietnam, Singapore, Japan
Cybersecurity for FinTech	Symantec Palo Alto Networks	Partnerships with cybersecurity firms in Singapore and Japan Collaborates with financial regulators to improve cyber-security standards across Asia's banking and FinTech sectors

Source Frana (2024), Kshetri (2020), Chodak (2024)

systems. This is furthermore likely to deepen divides in cybersecurity and data governance, as both regions seek to secure their technological infrastructures and protect their intellectual property. Adding the US and other Asian countries to the picture certainly intensifies this rivalry.

5 New Dynamics in Global Economic Governance

Few would argue that the post-Cold War era was mainly characterized by US unipolarity, with the US playing a dominant role in shaping the global economic order. The initial shift, which became even more evident after the GFC, was from a US-led unipolar system to what can be considered as a Sino-US bipolar system. A further phase or continuation of this shift manifested itself, perhaps before but surely after, the COVID-19 pandemic, which is a clear shift from West-centric to multilateral and multipolar governance. Countries like China, India, and other BRICS members are asserting more influence in global institutions and creating alternative platforms, such as the New Development Bank (NDB), to support development in emerging markets. While the G7 was previously the main group in global governance, the G20 now includes a more diverse set of economies, allowing emerging economies a larger role in shaping global agendas on trade, climate, and development (Slaughter, 2020). Challenges to such multilateralism have also arisen, such as:

- the fact that geoeconomics and the emphasis on relative gains are putting significant strain on institutions like the WTO;
- existing rules and dispute settlement mechanisms have become inadequate to address the complexities of the current geopolitical and economic landscape (Jayasuriya, 2020);
- deadlocks, the emergence of parallel informal associations like BRICS+ and attempts to bypass or undermine established institutions are becoming increasingly common; and
- the increase of 'contested multilateralism' in that rising powers like China are increasingly challenging the existing norms, rules, and practices of neoliberal multilateral institutions.

Even the dynamics of globalisation have changed since the 1990s if one compares the sheer dominance of the United States during that period (until the GFC) with today's multipolar geopolitical and -economic realities. Developments over especially the last decade confirmed that the international economic order is moving away from the neoliberal order that has dominated much of the post-Cold War era, moving instead toward a new "geoeconomic order" (Roberts et al., 2019: 655). This emerging order, marked by an increasing securitization of economic policy and economization of strategic policy, is likely to bring substantial changes to the rules, norms, and institutions governing global trade and investment.

Traditionally, the IMF, World Bank and WTO were the dominant role players in global economic governance. Specifically the first two of these institutions were primarily US-centric given that the US played a key role in establishing and maintaining these institutions, with their headquarters on US soil, not far from the White House in Washington, D.C. They often promote policies aligned with US economic interests (Wade, 2011). Particularly the dominance of the US dollar through its role as the world's reserve currency gave the US significant leverage in international finance. The IMF performs the role of ensuring stability in the international financial system (prioritizing monetary cooperation and economic growth) by: (1) providing policy advice to member countries; (2) monitoring countries' economic policies and their implications for global stability; (3) providing financial assistance by offering loans to countries experiencing economic difficulties or crises; and (4) providing technical assistance and training to help countries strengthen their economic institutions and improve their fiscal management and economic policies (IMF, 2024).

The World Bank's mandate is to reduce poverty and promote long-term economic development. By working together with its member countries, the World Bank aims to create a world free of poverty and a sustainable planet for all (World Bank, 2024). By providing loans and grants, offering technical assistance, and by mobilizing private capital, the World Bank seeks to build resilience, promote shared prosperity, and advance global goals (e.g., the SDGs). The WTO plays a crucial role in managing and facilitating global trade by promoting open, fair, and rules-based trading systems. It does this by administering trade agreements, resolving trade disputes, facilitating trade negotiations, ensuring global trade stability, and by monitoring trade policies. Although they play a vital role in the global economy, all three institutions of global economic governance face severe challenges due to both, internal rigidities and external changes in the global economic landscape. Some of the main concerns include:

- *Lack of democratic representation*: the voting power within these institutions is disproportionately held by wealthy nations. In the IMF and World Bank, the US and the EU countries have over 40% of the voting power while they represent only 10% of the global population. China and India, together, holds less than 10% of the voting power in these institutions, yet they represent almost a third of the global population. While the WTO operates on a one-country, one-vote system, the US and EU exert outsized influence through economic weight, trade volumes, and negotiation dominance.
- *One-size-fits-all approach*: the IMF's strict conditionalities attached to its loans often mandate austerity measures that worsen inequality and harm vulnerable populations. The World Bank's uniform approach to development projects is often viewed as ineffective or harmful to local communities and environments. The WTO's consensus-based decision-making has led to stalemates in trade negotiations (e.g., the stalled Doha Development Round). The institutions

mostly adopt standardized approaches to economic problems, failing to consider country-specific circumstances and challenges. Particularly the neoliberal policies that they promote have led to critics accusing them of Western bias.

- *Lack of transparency and accountability*: The decision-making processes within these institutions are quite opaque and not inclusive enough, which limit public scrutiny and accountability. One specific concern relates to their impact assessments where there is often a lack of rigorous evaluation of the social and environmental impacts of the policies and projects supported by these institutions. Persistent global inequalities and debt crises in the Global South have eroded trust in these institutions' ability to foster equitable development. Such failure to deliver results has affected their legitimacy negatively.
- *Failure to address modern global challenges*: while they prioritize it, these institutions have been slow to address the urgent issue of climate change and its impact on global economic stability. The same can be said of them adequately addressing the growing issue of inequality within and between countries. Despite their role in regulating the global financial system, they have been criticized for failing to prevent or mitigate financial crises. They have also struggled to regulate the digital economy and address issues like e-commerce, digital taxation, and intellectual property in the digital age. In terms of global power shifts, the institutions are perceived as unable to address geopolitical tensions such as US-China trade disputes or Russia's war in Ukraine, while struggling to adapt to the decoupling of global value chains and the rise of protectionism.

While the IMF, World Bank and the WTO to varying degrees of success have addressed some of these aspects, their reform is still too slow and inadequate. They are failing to evolve and address the complexities of the twenty-first century economy. The rise of new institutions in the East shows that countries want to move beyond the inadequacies and are seeking alternatives to the Bretton Woods institutions. Due to their declining relevance, countries increasingly bypass these institutions in favour of bilateral or regional agreements that better meet their interests. China and other Asian countries have initiated their own global economic institutions that are more aligned with their needs and aspirations. This has brought about a significant shift in the global economic governance arrangements of almost a century as they effectively challenge the dominance of existing US-led institutions. For instance, the Asian Infrastructure Investment Bank (AIIB) stands as a clear symbol of Asia's growing influence in the emerging global economic order. Operational since January 2016, it has already become the world's second-largest multilateral development institution, surpassing the Asian Development Bank (ADB) as the leading development bank in the Asia–Pacific region (Zhao, 2023). Based in Beijing and led by Jin Liqun, a former Chinese vice-minister of finance, the AIIB was established as a China-driven initiative. With top credit ratings from major agencies, the AIIB is seen as a potential rival or substitute to the IMF and World Bank.

The AIIB is a central part of China's regional infrastructure plan. In less than a decade, its membership has grown from 57 to 109. It now includes most of the G7 countries, 42 countries in Asia (excluding Japan), 22 in Africa, 26 in Europe, nine in Oceania, eight in South America, and two in North America (excluding the US and Mexico). Initially capitalized at US$100 billion, with China the largest contributor (US$50 billion), the AIIB is led by China, with India as the second-largest shareholder and main borrower (Raharyo, 2018). China's dominant 27% voting share, which is significantly larger than India's 8% and Russia's 6%, highlights the institution's Chinese-centric nature. Non-Asian members are limited to 25% of the shares and voting rights, ensuring Asian dominance. The AIIB, with the BRI as a central focus, can be seen as an international extension of China's new economic development model. This approach emphasizes long-term economic growth through systematic, widespread infrastructure investments (Javed et al., 2021). In contrast, institutions like the IMF, WTO, and World Bank promote Western neoclassical models focused on short-term export-driven and domestic consumption strategies. By shifting development funding and infrastructure investment away from Washington-centric dominance, the AIIB challenges the hegemony of the US dollar system in global financial flows.

The ADB consist of 68 member countries (49 regional and 19 non-regional members). Its purpose is to foster economic and social advancement in Asian and Pacific nations by supporting inclusive and sustainable development, offering financial resources, technical expertise, and policy guidance to its member states (ADB,). With a capital base of around $160 billion, the ADB mirrors the World Bank's weighted voting system, where voting power is linked to capital contributions. Based in the Philippines and historically led by Japanese presidents (with Japan being the largest shareholder), the ADB plays a crucial role in boosting demand and creating conditions for widespread privatization in sectors like transportation, urban development, energy, water, finance, and agriculture across the Asia–Pacific region (Ratnawatt, 2020). The ADB promotes corporate interests through public–private partnerships, loans, and co-financing arrangements, all underpinned by an infrastructure-driven growth strategy.

The New Development Bank (NDB), established in 2014, is a multilateral development bank that finances infrastructure and sustainable development projects in BRICS and other emerging economies. Governed by the principle of equality, the NDB is a joint effort of the BRICS nations, with no single member holding veto power. Also known as the 'BRICS Bank', it embodies the shared objectives of its members in fostering economic growth and cooperation. Its headquarters is in Shanghai, China, and was initiated by India at the 2012 BRICS summit (Wang, 2019). The five BRICS nations created a US$100 billion reserve currency pool, divided into 1 million shares having a par value of US$100 000 each.

In 2021, the NDB welcomed Egypt, Uruguay, Bangladesh, and the United Arab Emirates as new members. In May 2022, the NDB opened a regional office in Gujarat, India, aimed at financing and managing infrastructure projects in India and Bangladesh. Saudi Arabia announced its intention to join the NDB in May 2023. The NDB's ability to issue loans in local currencies (with more flexibility), rather than primarily relying on the US dollar as the IMF and World Bank do, challenges the dominance of the dollar in global trade and finance. It also offers a more inclusive and locally focused model of development finance. This alternative is particularly appealing to countries that want to reduce their dependence on the US dollar and the Western financial system. In this way, the NDB contributes to a more multipolar world, where emerging economies have greater influence in shaping global economic rules and norms.

While the long-term impact and the extent to which these new Asian institutions can truly challenge the existing global economic governance system remains to be seen, they are indeed reshaping the global economic landscape. The AIIB, ADB and NDB underscore the shift toward South-South cooperation, where developing countries in Asia, Africa, and Latin America are forming financial alliances and partnerships that provide an alternative to the Western-dominated systems. Yet, in the absence of consensus—or even lasting and thoughtful discussion—on what a new global order might look like, all the institutions of global economic governance (old and new) are currently under significant strain due to a combination of structural, geopolitical, and systemic challenges (Wang, 2019). They struggle to adapt to a multipolar world, regional rivalries, and evolving economic realities like digital transformation and climate change. Moreover, power has become central to the preservation, reform and formation of global economic institutions. Diversity and difference are now basic features of the global system as well as competition for power and contending conceptions of order between West and East.

The dynamics of global economic governance have also changed in terms of new governance and economic models. As Asian countries take on greater roles in institutions like the G20, WTO, and UN, they advocate for policies that align with their priorities, such as state sovereignty and economic development over Western-style political liberalism (Raharyo, 2018). The rise of these East-initiated institutions highlights alternative economic models that prioritize economic sovereignty, development autonomy, and non-interference in domestic affairs. These banks generally offer financing without the stringent conditionalities that Western institutions are often criticized for imposing. This diversification allows developing countries to choose financial and developmental partnerships that align with their own goals, potentially reducing reliance on traditional Western-dominated financial institutions.

One strong influence on the dynamics of global economic governance is competing ideologies, which is something the Bretton Woods institutions have not had to deal with before the GFC. The AIIB, NDB, and ADB offer an alternative to the Washington Consensus, which is often associated with the IMF and World Bank's market-driven reforms. The new East-driven financial institutions focus more on development-first policies, often with less emphasis on austerity measures and free-market liberalization, which have been controversial in many developing countries. In addition, a leading country like China advocates for 'statist/state-centric multilateralism,' which prioritizes state authority over individual rights and freedoms, and emphasizes the leadership of major states (Attiná & Feng,

Table 2 Key differences between Western and Eastern multilateralism

Aspect	Western model	Eastern model
Focus	Universal norms, liberal values	Pragmatic development, sovereignty
Governance	Weighted by financial contribution	Equitable power sharing
Conditionality	Stringent conditions on reforms	Minimal to no political conditions
Approach to cooperation	Standardized global framework; promote free trade, open markets, globalization/ integration as pathways to development; market-oriented frameworks	Tailored, region-specific solutions; prioritize practical outcomes over universal standards; state-led initiatives and stronger role for state-owned enterprises (SOEs)
Institutions	Dominated by legacy institutions (IMF, World Bank, WTO); hierarchical power	Rise of alternatives (AIIB, NDB, ADB, Shanghai Cooperation Organization (SCO)

Sources Woods (2023), Zhao (2023)

2023). This approach in many ways conflicts with the more liberal and universal-rules models of multilateralism favoured by Western nations. The extent of the differences is summarized in Table 2 below.

While these models are not strictly defined and there is overlap between them, the key distinctions are clear from the summary above. Notably, it is in the area of policy influence, which is highly diversified, where the differences appear to be most apparent. Asian-led alliances and forums, such as ASEAN and SCO, are shaping policies that prioritize regional stability, economic development, and sovereignty, countering Western influence (universal rules and liberal norms) and creating a more multipolar governance framework (Woods, 2023). These contrasting models reflect dissimilar historical experiences, values, geopolitical goals, and competing visions for global governance and shape the ongoing rebalancing of international power dynamics. A critical aspect in going forward is how the growing US-China rivalry is reshaping international cooperation. As China seeks global influence and the US aims to counterbalance China and reduce its international commitments, three key aspects of multilateralism are emerging: (1) managing superpower rivalry through formal and informal agreements and international institutions; (2) fostering cooperation beyond the immediate interests of powerful states; and (3) ensuring the legitimacy of international institutions. The evolution of these factors will determine the future of multilateralism.

A further significant impact on the changing dynamics of global economic governance is the evolving financial and monetary system. Some countries—especially those more aligned with China and the East—are seeking to reduce their reliance on the US dollar, leading to the emergence of alternative currencies and payment systems. The US dollar's dominance in global trade and finance is increasingly being challenged by other currencies, including the euro, Chinese yuan (renminbi), and cryptocurrencies. Even the IMF's Special Drawing Rights (SDR) are part of a more diversified reserve currency system that serves as an alternative to the dollar. China and Russia, for instance, are making concrete efforts to internationalize the yuan (e.g., through the Belt and Road Initiative and inclusion in the IMF's SDR basket) and the rouble to reduce reliance on the dollar (Heeks et al., 2024).

In addition, CBDCs and decentralized cryptocurrencies are transforming payment systems, cross-border transactions, and financial inclusion. Such digital currencies could reduce the dependency on traditional banking systems and alter global financial governance by bypassing intermediaries like SWIFT.[2] In fact, digital currencies and crypto-assets are disrupting traditional financial systems, raising questions about monetary policy, financial regulation, tax policy, and financial stability (Azar et al., 2022). From a geopolitical perspective, countries are vying to become leaders in the development and regulation of digital currencies, which could have a significant effect on altering the global balance of power. Concerns are that geopolitical tensions can lead to fragmentation of the global financial system, hindering international trade and investment.

Another factor in the changing dynamics of global economic governance is the fact that countries are increasingly imposing regulations on digital trade, data privacy, storage, and cybersecurity, with the EU's General Data Protection Regulation (GDPR) and China's Data Security Law serving as models. Countries are competing to shape global technology standards (e.g., for 5G, AI, and IoT), influencing who controls key industries and technologies (Skare & Soriano, 2021). It has created varying standards around data governance that is complicating international trade in digital services. Governing the digital economy and taking technological transformation at different levels (national and local) into account, has certainly become a growing challenge. This is part of the reason why CBDCs are being explored by countries as a way to maintain monetary sovereignty and influence cross-border payments. Multinational tech giants like Google, Alibaba, and Amazon have become key players in shaping the global economy, influencing trade, innovation, and regulation. They and many other multinational corporations (MNCs) possess significant economic and political power, which is challenging traditional notions of national sovereignty and corporate governance. Cities like Shenzhen, Bangalore, and Singapore are also emerging as global digital innovation centres, shifting economic power away from traditional industrial hubs in the West.

The US-China tech rivalry has led to efforts to 'decouple' technology ecosystems, with initiatives like the CHIPS Act and China's self-reliance strategies. This fragmentation affects supply chains, innovation, and global cooperation. In addition, the increase in cyber threats and espionage activities demand international collaboration to establish cybersecurity norms and protocols. Other aspects of the global nature of digital technologies are also posing challenges that require international cooperation. These include: the digital divide (inequality due to unequal access to digital technologies); AI displacing traditional jobs, and ethical AI use;

[2] SWIFT (Society for Worldwide Interbank Financial Telecommunication), a global financial messaging network founded in 1973, is owned and used by banks and other financial institutions to initiate international payments.

regulatory frameworks for distributed ledger technologies (blockchain); taxing digital transactions and multinational tech companies; and integrating digital tools into climate governance strategies (e.g., in achieving the SDGs).

It is clear that technological transformation and the digital economy are decentralizing economic power and redefining the parameters of global economic governance (Woods, 2023). These shifts demand new multilateral and regulatory frameworks, inclusive digital infrastructure development, and cooperative approaches to ensure that technological progress benefits all. Global governance institutions in both the East and the West will need to adapt to balance innovation with regulation to address global challenges in a rapidly digitizing world effectively.

Importantly, the latest and most significant shift is that the 'we' now involved in both geopolitics and global economic governance has changed in fundamental ways in that the diffusion of power and agency and the revolt against Western dominance are far broader than the rise of China. Attention has shifted toward non-Western 'rising powers,' social movements in the South, non-Western religious transnationalism (e.g., Islamic networks), and certainly, the ascent of China and the East. However, as part of a deeply rooted historical process, there has been a broad diffusion of political agency, enabling a much wider array of states, social groups, and societies to assert cultural recognition and press for changes in the legal and normative framework of global society (Hurrell, 2024). These growing influences are challenging traditional Western narratives in international relations and contributes to the cultural and normative pluralism of global society. It also underscores the influence of diverse perspectives on international issues, such as social justice, human rights, and community solidarity.

Lastly, there is a need for reconciling geopolitics and global economic governance in shaping a new, 'Geoeconomic Order'. The task for the future is not to choose between geopolitics and global governance, but to find ways to reconcile the two. Traditional geopolitical concerns, such as managing the balance of power, will need to coexist alongside complex forms of governance required to address global challenges like climate change and financial stability. This will require new approaches to diplomacy and international cooperation that can navigate the tensions between national interests and global imperatives.

Geoeconomics is certainly gaining in prominence, where states strategically utilize economic tools to achieve geopolitical objectives. This includes employing measures like sanctions, trade restrictions, supply chain control, investment constraints, and technological restrictions, not just to stimulate growth, but also to strengthen their geopolitical positions, protect key industries, and to exert influence and pressure on other nations (Roberts et al., 2019). It implies the 'securitization' of economic policy, where decisions on trade, investment, and resource allocation are influenced by concerns over sovereignty and strategic dominance, especially in areas like technology and energy. For example, countries are tightening FDI screening processes, particularly for sectors deemed critical to national security, such as defence, data governance, and critical infrastructure, adding layers of complexity to international investment (IMF, 2023). Geoeconomics also involves the 'economization' of strategic policy, where nations prioritize economic resilience and self-reliance to reduce vulnerabilities. This shift is visible in rising tariffs, sanctions, investment controls, and a focus on domestic industries, as well as efforts to reshape supply chains to lessen dependence on geopolitical rivals (Hurrell, 2024).

The growing use of economic statecraft has implications for global trade, investment flows, and the stability of the international financial system. A shift towards a geoeconomic order, while potentially disruptive, also opens up possibilities for innovation and reform. The shift toward a geoeconomic order represents a movement from a global economic landscape dominated by neoliberal ideals, such as free trade and open markets, toward one where economic policy is increasingly intertwined with national security and strategic objectives (e.g., security-related trade measures and industrial strategy). Finding effective ways to manage the interplay between geopolitics, geoeconomics, and geotechnology will be crucial for shaping a more stable and prosperous future. As part of the evolving dynamics in global economic governance, these are some of the most important institutional pressure points where geopolitics intersects with global economic governance and requires careful and collaborative management.

6 Conclusion

Henry Kissinger (2015) aptly described our current reality as a multipolar world shaped by contradictory realities, but these complexities go beyond a simple liberal-illiberal divide, offering opportunities amid challenges. Western liberals have historically oscillated between relying on unequal power structures, like empires or US hegemony, and rigidly pursuing balance-of-power politics based on fixed alliances and rivalries. The perceived shift in the global balance of power from West to East are not only causing a rethink of traditional power-relations due to increased fluidity, but also a repositioning with regards to new global alliances and shared interests.

It is important to note here that the key aspect of 'global' lies not so much in the nature of issues like climate change or interdependence but in the expanded capacity of diverse states and social actors to actively shape international law and society (Hurrell, 2024). This diffusion of agency and political consciousness is central to the globalization of international society, especially in the twenty-first century. A major factor today in this regard is the competition for technological dominance, which is reshaping the global geopolitical landscape as countries compete for leadership in areas like AI and 5G. The ripple effects of this on trade and investment flows is restructuring supply chains, regionalizing trade flows, and fuelling investment in emerging markets (Mirrlees, 2024). The technological innovation is positive, but in a climate of intense geopolitical rivalry, it comes at the cost of increasing global trade polarization, geoeconomic fragmentation, and targeted divestment.

Though it remains a topic of active debate, the rules-based system implemented by mainly the IMF, World Bank and WTO are still relevant in maintaining global economic stability. They do, however, face significant challenges in adapting to the demands of a rapidly evolving global landscape dealing with the realities of a multipolar, digital, and climate-conscious global economy. Their perceived inability to adapt and address the complexities of the economy in the twenty-first century—from technological shifts to environmental crises—has led to calls for reform or the development of alternative frameworks for global governance. The rise of alternative global economic governance institutions in the East, such as the AIIB, ADB, and NDB, are often seen as a way to address the perceived shortcomings of the Western-dominated system and to promote a more equitable and balanced global economic order. They do appear to be better suited to serve the needs of their member countries, which are mostly developing nations. This involves improved transparency, more democratic representation, greater willingness to innovate, and their ability to adopt a more flexible and nuanced approach to economic policy.

Instead of seeing it as a threat, this calls for strengthening international cooperation, institutional reform, and developing new mechanisms to address global economic challenges. It requires a commitment to multilateralism, a willingness to compromise and to innovate, and a focus on equitable and sustainable development. The analysis of historical and contemporary international cooperation suggests several ways to sustain cooperation, even amidst growing US-China rivalry (Woods, 2023). Global governance institutions can play a pivotal role by identifying areas where cooperation and competition can coexist and leveraging their capacities to share information, lower transaction costs, and broker and monitor agreements. Achieving this requires strong, high-quality leadership capable of forging agreements, persuading countries to pool resources, and managing an effective and motivated staff. Member states must actively define the criteria for effective institutional leadership, prioritize proven expertise in appointing leaders, and monitor their performance continuously. International institutions can enhance their legitimacy by ensuring inclusive participation, swiftly responding to global crises, and delivering essential global public goods in a changing economy. To navigate these changes effectively, policymakers, regulators, and international institutions must collaborate to develop a robust and resilient global financial architecture.

Global economic governance, for instance, is becoming more diversified, with emerging markets playing a larger role alongside traditional powers. This shift is reshaping the rules and institutions of global trade, investment, and development in response to new economic, technological, and environmental priorities. As power become more distributed across regions and institutions, this new multipolar approach could create opportunities for more inclusive growth, but it also brings the challenge of reconciling differing priorities and policy standards. It presents an alternative perspective on order, viewing it not as consensus or the elimination of contestation, but as the framework of rules and mutual understandings within which competition occurs. Order involves agreed boundaries for behaviour that avoid open hostility and explores the extent to which cooperation can continue in other areas, even amidst ongoing rivalries.

The transition from a Neoliberal Order to a Geoeconomic Order signifies a fundamental shift in the logic governing international economic relations. In view of increasing complexity, this change reflects an increasing concern among states that economic gains by one country may come at the expense of another, leading to a zero-sum mentality (Roberts et al., 2019). For example, China's economic policies, such as subsidies, investment screening, and data flow regulations, often clash with the principles upheld by Western nations like the US and members of the EU. This shift in focus complicates international cooperation and necessitates a reassessment of existing trade and investment rules (Hurrell, 2024). It introduces new dynamics in global economic governance, which requires more than institutional reform. Shared interests at state level and multinational level places pressure on global institutions in the East and West to find ways to not only promote collaboration, but also be agreement-mediators between states on opposite sides of the spectrum.

All these dynamics represent a significant departure from the post-Cold War era of globalization and pose critical questions about the future of global economic governance. There is a growing need for new frameworks and institutions that can address the complexities of geoeconomics, security concerns, and the shifting balance of power effectively. The following are some potential pathways and challenges that are worth noting:

- *The primacy of political order*: global economic governance currently requires a renewed focus on establishing and maintaining a stable political order. This involves addressing geopolitical rivalries, managing shifts in the balance of power, and strengthening core institutions like diplomacy and international law. Without a stable political foundation, efforts to build effective global economic governance are likely to be weakened.
- *The need for new institutions and norms*: current institutions and norms governing global economic relations are inadequate to address the challenges of the new economic order. New institutions and norms are needed that can better account for the interlinkages between security and economics, the diffusion of power, and the rise of geoeconomics. This should involve reforms to institutions like the WTO, IMF and World Bank, or the creation of new institutions tailored to specific challenges, such as cybersecurity and climate change.
- *The challenge of legitimacy*: as global governance evolves, a key challenge will be ensuring the legitimacy of these new institutions and norms. It will be crucial to incorporate the perspectives and interests of a wider range of actors, notably those from the Global South, who may have different conceptions of fairness and justice. Building legitimacy will require greater transparency, inclusivity, and accountability in decision-making processes.
- *Reconciling geopolitics and global governance*: the task for the future is not to choose between geopolitics and global governance, but to find ways to reconcile the two. Traditional geopolitical concerns, such as managing the balance of

power, will need to coexist alongside complex forms of governance required to address global challenges like climate change, inequality and financial stability. This will require new approaches to diplomacy and international cooperation that can navigate the tensions between national interests and global imperatives.

Clearly, the economic implications of the possible geopolitical shift from West to East are vast, substantial, and evolving by the day. The role of global economic governance in steering the way are becoming more and more important, which requires change beyond mere global institutional reform. A complete reformulation of global economic governance arrangements might be required, as pointed out in this chapter. For the immediate future, there is the unlikely possibility of the full restoration of US hegemony in and after the new Trump era or restored Western hegemony, but the more likely scenario is the move to a rather balanced multilateral international economic order.

This is even more likely in view of the instability in France and Germany. Both governments collapsed in 2024. Former French Prime Minister Michel Barnier lost a no-confidence vote in parliament; this after President Macron held snap elections (which his party lost) in July to elect a new National Assembly. France, being Europe's only large military power along with the UK (who also had a dramatic change of Prime Minister in 2024), is in turmoil for economic reasons. It is the second largest economy in the Eurozone, with a ballooning budget (way beyond EU norms), escalating debt, and worried taxpayers as living costs keep rising. Germany, the EU's largest economy, is in serious trouble after it shrank 5% compared to pre-pandemic levels due to high energy costs, falling exports, and rapid deindustrialisation.

The economic model underpinning Europe's economy seems to be declining fast. German households lost up to €2500 annually and the auto industry risks losing 40% of its value in the last decade (Adler, 2024). Germany needed to hold a snap election after its bickering coalition government collapsed—being at loggerheads over how to revive the German economy. The fact that France and Germany, who are traditionally seen as the main drivers of Europe's ideological and political horsepower, are not showing strength and unity is leaving the EU bloc directionless and weakened. Trump's return to the White House actually fills the EU and Europe with geopolitical fear, given the US trade deficit with Europe and, according to Trump, Europe's insufficient defence spending. This does not bode well for a strong and unified West.

As Hurrell (2024: 222) puts it, "the future is unlikely to be about the replacement of one global hegemon by another". The effect of a potential shift in the balance of power towards the East in fact brings global power closer to equilibrium rather than hegemony. The new status quo might be more determined by countries' responses to, instead of the actual actions of, the US and China as the two main poles of power. Ironically, the current period of intensified geopolitical risk, as part of an evolving global economic order, could usher in some form of 'forced' cooperation between East and West given the renewed centrality and existential threat of nuclear weapons on the world stage. A prolonged and uneasy period of coexistence is likely, where states with vastly different domestic economic systems and political values remain deeply economically interconnected. This interdependence may be further reinforced by their shared need to address critical global sustainability challenges, such as climate change, geopolitical instability, and the governance of emerging technologies.

References

ADB (Asian Development Bank) (2016a). *ADB Annual Report 2016*. ADB

ADB (Asian Development Bank). (2016b). *ADB Annual Report 2016*. ADB.

ADB (Asian Development Bank). (2023). *ASEAN and global value chains: Locking in resilience and sustainability*. ADB.

Adler, K. (2024). Why France's turmoil is grave concern for Europe. *BBC*. https://www.bbc.com/news/articles/c0q0zvvjj19o

Aljazeera. (2023). *China's Xi tells Putin of 'changes not seen for 100 years'*. https://www.aljazeera.com/news/2023/3/22/xi-tells-putin-of-changes-not-seen-for-100

Attiná, F., & Feng, Y. (2023). *China and world politics in transition: How China transforms the world political order*. Springer.

Australian Government. (2022). *Regional Comprehensive Economic Partnership Agreement (RCEP)*. Department of Foreign Affairs and Trade. https://www.dfat.gov.au/trade/agreements/in-force/rcep

Azar, P. D., Baughman, G., Carapella, F., Gerszten, J., Lubis, A., Perez-Sangimino, J. P., Rappoport, D. E., Scotti, C., Swem, N., Vardoulakis, A., & Werman, A. (2022). *The financial stability implications of digital assets*. Staff Reports, 1034, Federal Reserve Bank of New York.

Banach, C., & Gunter, J. (2023). *How the BRI is shaping global trade and what to expect from the initiative in its second decade*. MERICS China Global Competition Tracker No. 4. https://merics.org/en/tracker/how-bri-shaping-global-trade-and-what-expect-initiative-its-second-decade

Bajpaee, C. (2023). Reinvigorating India's 'Act East' Policy in an age of renewed power politics. *The Pacific Review, 36*(3), 631–661.

Beker, V. A. (2014). The European debt crisis: Causes and consequences. *Journal of Stock & Forex Trading, 3*(2), 1–9. https://doi.org/10.4172/2168-9458.1000115

Bordo, M.D. (2017). *The operation and demise of the Bretton Woods system: 1958–1971. NBER Working Paper series*, 23189, pp. 1–37.

Buchholz, K. (2024). China leads the global e-commerce market. *Statista*. https://www.statista.com/

Budanov, K., Vereshchak, V., Kudriavtsev, V., Mokliak, S., & Rubel, K. (2024). Causes and trends in modern geopolitical changes. *Edelweiss Applied Science and Technology, 8*(6), 2261–2269.

Carew, A. (1987). *Labour under the Marshall Plan: The politics of productivity and the marketing of management science*. Manchester University Press.

Chodak, G. (2024). *Current and new payment methods in e-commerce: Innovations and developments*. Springer.

Dabla-Norris, E., Kinda, T., Chahande, K., Chai, H., Chen, Y., De Stefani, A., Kido, Y., Qi, F., & Sollaci, A. (2023). *Accelerating innovation and digitalization in Asia to boost productivity*. IMF.

De Marcos, P. (2022). *Impact of the Belt and Road Initiative on bilateral trade with China*. Thesis advised by Professor Barry Eichengreen in the Department of Economics at the University of California, Berkley.

Eichengreen, B. (2021). Bretton Woods after 50. *Review of Political Economy, 33*(4), 552–569. https://doi.org/10.1080/09538259.2021.1952011

Emeagwali, G. (2011). The neo-liberal agenda and the IMF/World Bank structural adjustment programs with reference to Africa. In: Kapoor, D. (Ed.), *Critical perspectives on neoliberal globalization, development and education in Africa and Asia*. Brill.

Erie, M. S., & Streinz, T. (2021). The Beijing effect: China's digital Silk Road and transnational data governance. *Journal of International Law and Politics, 54*(1), 1–92.

Frana, P. L. (2024). Assessing smart nation Singapore as an international model for AI responsibility. *International Journal on Responsibility, 7*(1), 1–30.

Gaofeng, Y., Krishna, V. V., Zhang, X., & Jiang, Y. (2021). Chinese Universities in the National Innovation System: Academic Entrepreneurship and Ecosystem. *Routledge*. https://doi.org/10.4324/9781003015536

Geeraerts, G. (2011). China, the EU, and the new multipolarity. *European Review, 19*(1), 57–67. https://doi.org/10.1017/S1062798710000335

Gökay, B. (2021). Two pillars of US global hegemony: Middle Eastern oil and the petrodollar. In I. Ness & Z. Cope (Eds.), *The Palgrave encyclopedia of imperialism and anti-imperialism*. Palgrave Macmillan. https://doi.org/10.1007/978-3-030-29901-9_334

Hardcastle, D., Tsukamoto, Y., Tan, K., Wong Harris, T., Park, K., & Lee, B. (2024). *Southeast Asia's Green Economy 2024 Report: Moving the needle*. Bain & Company, GenZero, Standard Chartered, and Temasek.

Hassan, T. A., Hollander, S., Van Lent, L., & Tahoun, A. (2019). Firm-level political risk: Measurement and effects. *Quarterly Journal of Economics, 134*(4), 2135–2202. https://doi.org/10.1093/qje/qjz021

Hayashi, D. (2020). Harnessing innovation policy for industrial decarbonization: Capabilities and manufacturing in the wind and solar power sectors of China and India. *Energy Research & Social Science, 70*, 1–16. https://doi.org/10.1016/j.erss.2020.101644

Heeks, R., Ospina, A. V., Foster, C., Gao, P., Han, X., & Jepson, N. (2024). China's digital expansion in the Global South: Systematic literature review and future research agenda. *Information Society, 40*(2), 69–95.

He, S., & Guo, K. (2019). Examination of the energy trading status of China and India and the prospect for cooperation. *Procedia Computer Science, 162*, 819–826.

Hillman, J. E. (2021). Competing with China's Digital Silk Road. *Center for Strategic & International Studies*. https://www.csis.org/analysis/competing-chinas-digital-silk-road

Hurrell, A. (2024). Geopolitics and global economic governance. *Oxford Review of Economic Policy, 40*(2), 220–233. https://doi.org/10.1093/oxrep/grae013

Hu, Z., Liu, C., & Lee, P. T. (2021). China's global investment and maritime flows in the context of the Belt and Road Initiative. *Journal of Contemporary China, 30*(129), 465–480. https://doi.org/10.1080/10670564.2020.1827357

International Monetary Fund (IMF). (2023). *World economic outlook: A rocky recovery*. IMF.

International Monetary Fund (IMF). (2024). *World economic outlook (October 2024)*. IMF.

Javed, S., Bo, Y., Tao, L., & Dong, W. (2021). The 'dual circulation' development model of China: Background and insights. *Rajagiri Management Journal, 17*(1), 2–20.

Jayasuriya, K. (2020). The age of political disincorporation: geo-capitalist conflict and the politics of authoritarian statism. *Journal of Contemporary Asia, 53*(1), 165–178. https://doi.org/10.1080/00472336.2022.2092535

JETRO Investment Japan Report. (2023). Trends in greenfield investment in Japan. https://www.jetro.go.jp/en/invest/investment_environment/ijre/report2023/ch2/sec4.html

Jones, M. (2023). Top 9 tech hubs in Asia for 2023. Emerald Technology. https://blog.emerald-technology.com/top-tech-hubs-in-asia

Joshipura, M., Mathur, S., & Kedia, N. (2024). Sustainable investing and financing for sustainable development: A hybrid review. *Sustainable Development, 32*, 4469–4485.

Kshetri, N. (2020). China's emergence as the global fintech capital and implications for Southeast Asia. *Asia Policy, 15*(1), 61–82.

Kim, M. S., Kim, J., & Kim, S. (2023). Korea's leadership in 5G and beyond: Footprints and futures. *Telecommunications Policy, 47*(8), 1–11. https://doi.org/10.1016/j.telpol.2023.102613

Kissinger, H. (2015). *World order: Reflections on the character and course of history*. Penguin.

Kugelman, M. (2024). India's balancing act with the West as Brics flexes new muscles. *BBC*. https://www.bbc.com/news/articles/cly

Lee, C. (2024) China backs Russia-proposed BRICS payment system. *Voice of America*. https://www.voanews.com/a/china-backs-russia-proposed-brics-payment-system-/7836865.html

Lee, C., Strong, R., & Dooley, K. E. (2021). Analyzing precision agriculture adoption across the globe: A systematic review of scholarship from 1999–2020. *Sustainability, 13*, 1–15.

Lewis, J. I. (2023). *Cooperating for the climate: Learning from international partnerships in China's clean energy sector*. MIT Press.

Lin, C. (1984). Global pattern of energy consumption before and after the 1974 oil crisis. *Economic Development and Cultural Change, 32*(4), 781–802. https://doi.org/10.1086/451426

Mahoney, J. G. (2023). China's rise as an advanced technological society and the rise of digital orientalism. *Journal of Chinese Political Science, 28*, 1–24.

McKinsey & Company. (2022). *Decarbonizing Asia's supply chains*. https://www.mckinsey.com/featured-insights/future-of-asia/videos/decarbonizing-asias-supply-chains

Ministry of External Affairs. (2024). *Kazan Declaration: Strengthening multilateralism for just global development and security*. https://www.mea.gov.in/bilateral-documents.htm?dtl/38450/Kazan_Declaration__Strengthening_Multilateralism_For_Just_Global__Development_And_Security

Mirrlees, T. (2024). The US and China's digital tech war—A new rivalry within and beyond the US empire. *Socialist Register*, January, pp. 1–36.

Mirza, M., Naeem, M. A., Nguyen, T. T. H., Arfaoui, N., & Oliyide, J. A. (2023). Are sustainable investments interdependent? The international evidence. *Economic Modelling, 119*, 1–19. https://doi.org/10.1016/j.econmod.2022.106120

Naidu, G. V. C. (2006). India and the Asia-Pacific: The look East Policy. *Indian Foreign Affairs Journal, 1*(1), 89–103.

Nedelcu, M. (2024) The rise of geotechnology: A paradigm shift from geopolitics and geoeconomics in the context of Industry 4.0. *Sciendo*, 1979–1988. https://doi.org/10.2478/picbe-2024-0167

Pandey, B. (2022). The race to the starting line: The rise of CBDCs. *Chicago Policy Review*, 1–6.

Qiu, H., Xia, D., & Yetman, J. (2024). Deconstructing global trade: the role of geopolitical alignment. *BIS Quarterly Review*, 35–50.

Raharyo, A. (2018). Asian Infrastructure Investment Bank (AIIB) as an instrument for regional leadership of China in the East Asia region. *AEGIS Journal of International Relations, 2*(2), 109–124.

Ratnawatt, K. (2020). The impact of financial inclusion on economic growth, poverty, income inequality, and financial stability in Asia. *Journal of Asian Finance, Economics and Business, 7*(10), 73–85.

Roberts, A., Moraes, H. C., & Ferguson, V. (2019). Toward a geoeconomic order in international trade and investment. *Journal of International Economic Law, 22*, 655–676.

Rosenberg, A., & Sandrini, F. (2023). Geopolitical shifts and investment implications. *Amundi Institute*. https://research-center.amundi.com/files/nuxeo/dl/3df6c9a7-9e25-4968-b38c-66907f8cd40c?inline=

Sahakyan, M. D. (2020). Rebuilding interconnections: Russia, India and the International North-South Transport Corridor. *AsiaGlobal Online* 1–8.

Schindler, S., Alami, I., DiCarlo, J., Jepson, N., Rolf, S., Kemal Bayırbağ, M., Cyuzuzo, L., DeBoom, M., Farahani, A. F., Liu, I. T., McNicol, H., Miao, J. T., Nock, P., Teri, G., Facundo Vila Seoane, M., Ward, K., Zajontz, T., & Zhao, Y. (2024). The second Cold War: US-China competition for centrality in infrastructure, digital, production, and finance networks. *Geopolitics, 29*(4), 1083–1120. https://doi.org/10.1080/14650045.2023.2253432

Schumacher, D. (1985). The 1973 oil crisis and its aftermath. In: *Energy: Crisis or opportunity?* Palgrave. https://doi.org/10.1007/978-1-349-17797-4_2

Shaheen, I. (2013). South Asian Association for Regional Cooperation (SAARC): Its role, hurdles and prospects. *IOSR Journal of Humanities and Social Science, 15*(6), 1–9.

Skaf, Y. (2023). Cryptocurrencies and blockchain technology applications. In S. H. Jafar, K. Hemachandran, H. El-Chaarani, S. Moturi, & N. Gupta (Eds.), *Artificial intelligence for capital markets*. Taylor & Francis.

Skare, M., & Soriano, D. R. (2021). How globalization is changing digital technology adoption: An international perspective. *Journal of Innovation & Knowledge, 6*(4), 222–233.

Slaughter, S. (2020). *The power of the G20: The politics of legitimacy in global governance.* Routledge.

Smith, A. (2013). Europe and an inter-dependent world: Uneven geo-economic and geo-political developments. *European Urban and Regional Studies, 20*(1), 3–13. https://doi.org/10.1177/0969776412463309

Song, W., & Ai, W. (2023). China's vision for a future world order and its implications for global governance. In F. Attinà, & Y. Feng (Eds.), *China and world politics in transition. Global power shift*. Springer. https://doi.org/10.1007/978-3-031-27358-2_2

Stec, G. (2020). Towards clear-eyed R&I cooperation. *Friends of Europe.* https://www.friendsofeurope.org/insights/technology-views-from-east-and-west/

United Nations Trade and Development (UNCTAD). (2024). *World Investment Report 2024.* UNCTAD.

Vinokurov, E. Y., Ahunbaev, A., & Zaboev, A. I. (2022). International North-South transport corridor: Boosting Russia's "pivot to the South" and Trans-Eurasian connectivity. *Russian Journal of Economics, 8*(2), 159–173.

Wade, R. (2011). Emerging world order? From multipolarity to multilateralism in the G20, the World Bank, and the IMF. *Politics & Society, 39*(3), 347–378.

Walton, J. (2024). The 5 countries that produce the most solar energy. *Investopedia.* https://www.investopedia.com/articles/investing/092815/5-countries-produce-most-solar-energy.asp#:~:text=China%20leads%20the%20world%20as,of%20measurement%20of%20electrical%20power

Wang, H. (2019). The New Development Bank and the Asian Infrastructure Investment Bank: China's ambiguous approach to global financial governance. *Development and Change, 50*(1), 221–244.

Wang, C. N. (2024). *China Belt and Road Initiative (BRI) Investment Report 2023.* https://greenfdc.org/china-belt-and-road-initiative-bri-investment-report-2023/?cookie-state-change=1729934837504

World Bank. (2018). *Belt and road initiative.* https://www.worldbank.org/en/topic/regional-integration/brief/belt-and-road-initiative

World Bank. (2024). *Who we are.* https://www.worldbank.org/ext/en/who-we-are

World Trade Organization (WTO). (2023). *World trade statistical review 2023.* WTO.

Woods, N. (2023). Multilateralism in the 21st century. *Global Perspectives, 4*(1), 1–9. https://doi.org/10.1525/gp.2023.68310

Zhang, A., Alvi, M. F., Gong, Y., & Wang, J. X. (2022). Overcoming barriers to supply chain decarbonization: Case studies of first movers. *Resources, Conservation and Recycling, 186*, 2–12. https://doi.org/10.1016/j.resconrec.2022.106536

Zhang, J., Ziying, S., & Wang, C. N. (2024). New report: China green finance status and trends 2023–2024. *Green Finance and Development Center.* https://greenfdc.org/china-green-finance-status-and-trends-2023-2024/#:~:text=Green%20insurance%20lacks%20comparable%20data,in%20navigating%20market%20interest%20rates

Zhao, S. (2023). *The dragon roars back: Transformational leaders and dynamics of Chinese foreign policy*. Stanford University Press.

Zhu, K. (2024). *Ranked: The 15 countries with the most solar power installed.* https://www.visualcapitalist.com/countries-by-solar-power-ranking/

2 A Challenge of Sustainability: The US Economy, De-dollarization and the BRICS Plan

Abstract

This chapter investigates the multifaceted challenges to the US economy's sustainability, set against the backdrop of a shifting global economic order. It examines the US's economic vulnerabilities, including fiscal unsustainability, troubling trade deficits, income inequality, environmental issues, and its heavy reliance on the dollar as the global reserve currency amidst rising geopolitical tensions. The chapter finds that the US economy's vulnerabilities, including its high national debt and persistent trade imbalances, pose significant risks to global economic stability. It also identifies a strategic move towards de-dollarization by Eastern nations, particularly the BRICS+ alliance, to counter these vulnerabilities and challenge the US's economic hegemony. The significance of this chapter lies in its analysis of the intricate relationship between economic sustainability and geopolitical influence in an increasingly multipolar world. It underscores the potential reshaping of the global financial system and the urgent need for the US to adopt sustainable economic practices to maintain its global standing.

Keywords

Economic sustainability • US economy • De-dollarization • BRICS • SDGs

A. J. van Niekerk, *West to East: A New Global Economy in the Making?*,
https://doi.org/10.1007/978-3-031-93267-0_2

1 Introduction

Economic sustainability refers to productive actions/practices that support long-term economic growth without negatively affecting social development and the environment (OECD, 2024). It means creating a balance between GDP growth, social equity, and environmental protection so that growth is inclusive and ecologically sound. It involves ensuring that current economic activities do not compromise the resources or opportunities available for future generations to achieve prosperity and well-being. The US economy faces significant challenges concerning economic sustainability. Given that it is currently the largest economy in the world, its perceived lack of sustainability puts the global economy as a whole at risk. Global financial instability is one key implication whereby, for example, a US debt crisis could trigger a global financial crisis given that the US dollar is the world's primary reserve currency. So also, can economic instability in the US lead to increased market volatility, affecting global stock markets and currency exchange rates. At present, global economic sustainability depend strongly on US economic sustainability.

This is why countries in especially the East have started to diversify trade partnerships, develop alternative regional financial mechanisms through the AIIB and NDB, and are strengthening intra-regional integration. China and India have been at the forefront of reducing the risk posed by overreliance on the US economy by improving their own resilience and sustainability through investing in research and development (especially technology) and investing in other resilient economies to diversify and strengthen supply chains. Eastern countries are actively promoting local currencies in trade (limiting exposure to US monetary policies); they are building strategic reserves of foreign currencies, gold, and other assets to protect against potential US economic instability; and they are building strong partnerships with other emerging economies to help mitigate risks associated with the US economy (Mahoney, 2023).

In fact, across the globe—not just in the East—countries are adopting similar strategies to reduce their reliance on the US economy and to foster greater sustainability. Eurozone countries have been working to enhance the role of the euro in international trade and finance to reduce dependence on the US dollar. Initiatives like Mercosur in Latin America and partnerships with China emphasize diversification of trade away from the US. African nations are increasingly working with Eastern countries like China and India through initiatives like the Forum on China-Africa Cooperation (FOCAC). Gulf states in the Middle East are transitioning to diversified economies through initiatives like Saudi Arabia's Vision 2030 to lessen reliance on oil exports and the dollar-dominated global market.

Globally, economic sustainability has long been a challenge and an ideal the world economy is trying to work towards—at least in terms of the policy goals of governments and international institutions (e.g., the UN's SDGs). Balancing short-term gains (growth and profits) with long-term sustainability has always been problematic. Ensuring that economic growth benefits all segments of society (social equity), especially in the face of unequal development, remains a challenge. Due to global economic interconnectedness, a financial crisis in one region tends to have ripple effects on others, making it difficult to address sustainability issues in isolation.

Sustainability goals, such as reducing poverty, equitable resource distribution while promoting efficient use of resources to reduce environmental impact, investing in green technologies and sustainable industries to create new jobs, inclusive growth that can lead to greater social stability, long-term economic prosperity, and improved global cooperation, are all part of the effort to build more resilient and equitable economies (Joshipura et al., 2024). This shows the complex and multifaceted nature of economic sustainability. Although it seems like an elusive ideal, at least most countries around the world are prioritizing economic sustainability. How they approach it in the West and the East vary significantly, with many undercurrents of self-interest, geopolitical tensions, and global ambitions.

Given that the US economy plays such a decisive role in the global economy—in terms of both its strong influence and being a challenge to China and others—it is vital to examine the factors that make the US economy unsustainable. This is the focus of the next section. After that, the efforts towards de-dollarization is investigated as well as other sides and aspects related to it. For instance, the increasing use of financial sanctions as a geopolitical tool, as seen in the case of Russia, has raised concerns about the 'weaponization' of the US dollar and banking system. This is prompting some countries to seek alternatives to dollar reserves and payment systems, which could lead to a gradual diversification of international reserves and financial systems away from the dollar and towards other currencies, such as the renminbi. The section that follows this examines whether BRICS+ presents a challenge to US hegemony. The aim of the chapter is to, from a sustainability perspective, gain more clarity on the role of the US economy in a changing global landscape.

2 What Makes the US Economy Unsustainable?

First of all, a strong US economy depends on fiscal health. It is no secret that the US remains on an unsustainable fiscal path with national debt already at its highest level since World War II. Having surpassed 100% in 2013, its debt-to-GDP ratio rose to almost 130% in 2024, exceeding US$33 trillion (U.S. Treasury, 2024). This implies large and growing interest payments that transfer purchasing power to foreigners and reduce the potential growth of domestic demand. While some debt increase can be attributed to policy changes and underestimates in entitlement program spending, the primary cause of the rising US debt-to-GDP ratio is the

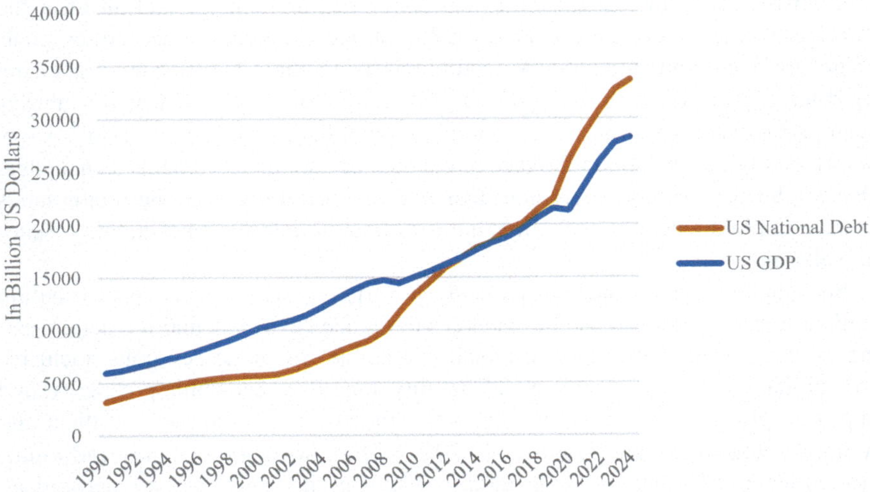

Fig. 1 US national debt compared to GDP (1990–2024). *Sources* Data from U.S Treasury (2024), Federal Reserve Bank of St. Louis (2024)

persistent gap between government spending and revenue, a trend that has been observed for several years (IMF, 2023). In 2023, the US fiscal deficit was US$1.7 trillion; the fourth year in a row that the deficit was over US$1 trillion. Figure 1 illustrates the acceleration of the growth rate of US national debt, especially after the GFC. It now clearly exceeds growth in GDP at an alarming rate.

The US government has consistently been unable to produce reliable financial statements due to accounting and management issues across several agencies, including the Departments of Defence and Education, the Small Business Administration, and Medicare and Social Security. This has limited the government from effectively addressing its financial challenges. There are mainly three contributing factors to the sharp rise in US debt: (1) changing demographics as society is aging, placing more pressure on the federal budget; (2) US healthcare costs are rising disproportionately compared to that in other advanced countries like Switzerland, Germany, and France; and (3) inadequate revenues resulting from federal spending (US$6.1 trillion in 2023) exceeding federal revenues (US$4.4 trillion in 2023), causing annual deficits and an increasing national debt balance (Srinivasan, 2024). The implications of the continued rise in US national debt include:

- *Higher interest payments*: puts a strain on government budgets and limit spending on essential services like education, healthcare, and infrastructure, plus higher interest rates limits capital access for businesses, reducing innovation, productivity, and growth.

- *Crowding out private investment*: excessive government borrowing can crowd out private investment, as investors may prefer to invest in lower-risk government bonds rather than riskier private ventures, which can generate higher returns.
- *Reduced fiscal flexibility*: limits the government's ability to respond to economic crises, natural disasters, or unexpected challenges, and reducing adaptability in future policy-making, which may force the implementation of austerity measures to reduce spending and raise taxes (Bergstresser, 2024).
- *Intergenerational burden and erosion of confidence*: future generations may face higher taxes or reduced public services to address accumulated debt, affecting their economic mobility and quality of life, and it may lead to reduced domestic and international confidence in the US economy, so increasing the risk of capital flight and market instability.
- *Currency depreciation*: high and persistent debt can lead to currency depreciation, which can increase inflation, reduce US geopolitical leverage, and decrease the purchasing power of US consumers (e.g. a weaker dollar can cause higher import prices and borrowing costs).

To address these challenges, it is crucial for the US to implement sustainable fiscal policies like reducing spending, increasing revenue, and growing its economy. Without effective management of national debt, the US will struggle to sustain economic stability, encourage investment, and maintain the nation's long-term economic resilience. Inclusive economic growth can strengthen the US's social fabric, reduce political polarization, and enhance its long-term economic prospects. It can help address systemic inequalities and foster broad-based prosperity. A more equitable economy promotes stability, which can mitigate the risks associated with economic shocks and financial crises. With the new Trump administration now in office, the US is unlikely to prioritize inclusive growth, given his administration's historical focus on deregulation, tax cuts, and limited social spending. The US economy also relies heavily on consumer spending, fuelled by credit rather than income growth, which can create vulnerabilities, especially during economic downturns. In 2024, the average debt per person excluding mortgages was US$22,713. When including all forms of debt (e.g., mortgages, personal loans, auto loans, credit cards, and student loans) the average debt per individual rises significantly. Recent data indicates that the average American consumer's total debt is around US$104,215 (Matthews, 2024; Warren, 2024). It is particularly the US's short-term focus—prioritizing immediate gains over long-term sustainability—that is its Achilles heel, making efforts to deal with its debt problem much more challenging.

A second major area of concern regarding the US economy's sustainability is its large trade deficit. Persistent trade deficits (US imports continuing to exceed exports, except for petroleum products and services) is leading to a reliance on foreign capital to finance its consumption-driven economy (Bolt & Van Zanden, 2024). The US has had a current account deficit of more than 1% of GDP in almost every year since 1981. In 2006, the deficit was 5.7% of GDP and remained high (in 2022 it was 3.77% of GDP and nearly US$ 1 trillion), while for most of the past decade, it has had a negative balance of payments. For half a decade, the US has experienced a decline in industrial dominance and business confidence, which contributed to the widening of its trade deficit.

Attempts to reduce reliance on Chinese imports through tariffs have yielded limited success. Despite some benefits from low-interest rates and income from capital invested abroad, the persistent trade deficit remains a concern. It is expected that the trade imbalance will persist, and the household and corporate sectors will continue to deleverage rather than increase spending. The US, being a sovereign country issuing a reserve currency, enjoys the advantage of running large current account deficits, which has been helped by low-interest rates and income from foreign investments to mitigate the impact of these deficits. However, a continued reliance on this 'exorbitant privilege' may now result in negative long-term consequences for the US economy.

Trade deficits can furthermore reduce national savings, which limit investment in domestic industries and infrastructure. The low American national savings rate has long been a serious concern to economic policymakers. From 1947 to 2024, total US net savings averaged at 6.3% of Gross Domestic Income (GDI), which is notably lower than that of many other advanced and emerging economies. The average savings rate in Germany, France, and the UK is between 10 and 15%, while in China it is over 30%. Bringing it into perspective is that between 2002 and 2024, the average rate in the US was 1.8% and nearly one in five quarters saw a negative net savings rate (Deller, 2024).

Furthermore, as the GDI increased from 2023 to 2024, savings actually decreased. This is due to an American culture of consumption, easy access to credit, and a strong social safety net. The US tried to address its savings shortfall by borrowing from abroad, enabling it to invest more than it saves over the past few decades. This reliance on persistent current account deficits has resulted in an increasing level of indebtedness to other countries, reflected in net US liabilities to foreign asset holders. However, many of these foreign nations financing US investment also face aging populations and declining national savings. Consequently, depending heavily on foreign savers to fund a significant portion of domestic investment is not a sustainable long-term strategy. In the short term, a negative net savings rate adds significantly to the negative impacts of a fiscal deficit and a trade deficit, plus it increases the vulnerability of the US economy to external shocks.

The US trade deficit reflects an over-reliance on imports, undermining its domestic production. Sectors unable to compete with cheaper foreign goods have shrunk in the last four years, resulting in job losses and reduced industrial capacity (Dorn & Levell, 2024). Trade deficits affect areas reliant on manufacturing and export-oriented industries disproportionately, leading to economic stagnation and inequality in those regions. The benefits of cheap imports have favoured consumers and capital owners in the US, while workers in affected industries have endured the worst due to job losses and wage stagnation. The United States' persistent trade deficit has weakened the dollar over time, as it spends more on imports than it earns from exports, which has led to growing fears about inflation and the erosion of purchasing power. Its dependence on foreign inflows to balance deficits increases its vulnerability to currency fluctuations and speculative attacks. The global imbalance caused by the large US trade deficit and the corresponding current account surpluses held by other countries, particularly in China, East Asia, and oil-exporting nations, is a major concern to the global economy, as these surplus economies finance US deficits by purchasing US Treasury bonds or other financial assets, creating a loop of dependency.

The US dollar's status as the global reserve currency exacerbates this imbalance. Countries prefer to hold US dollars and dollar-denominated assets, which allows the US to borrow cheaply abroad and sustain deficits without immediate repercussions. However, this increases global dependence on the stability of the US economy and monetary policy (Evgenidis & Malliaris, 2023). The economic risk is that over-reliance on the US economy creates vulnerabilities for surplus countries if US demand or economic stability falters. Many surplus economies like China, Japan and Germany, rely heavily on exports to the US. This dependency entrenches the global imbalance, as their economies are structured to maintain trade surpluses.

In addition, the US absorbs a large portion of global savings through foreign investments in its capital markets. This leads to a global imbalance where the US runs capital account surpluses in its balance of payments to try to offset its current account deficits. Hence, this global imbalance has implications for both the US and the global economy. For the US, it can lead to increased debt, currency depreciation, financial volatility, and reduced economic competitiveness. For the global economy, it can create imbalances in trade flows, currency exchange rates, and global financial stability. What is more, disputes over trade practices, currency manipulation, and economic policies add to the geopolitical tension, particularly between the US and major surplus nations like China. This global imbalance, driven by the large US trade deficit, underlines severe systemic challenges within the global economic architecture, where disparities in savings, consumption and trade patterns create significant economic and political vulnerabilities worldwide.

What thirdly makes the US economy unsustainable is growing concerns about income inequality. The 'American Dream' is rooted in the idea that hard work can lead to upward mobility. However, high levels of inequality often correlate with reduced social mobility, meaning individuals born into lower-income families face greater barriers to improving their economic status. For decades, liberal economic policies—originating from the West—have dominated policy thinking, not only in the US, but also globally. The liberal economic approach has often downplayed the significance of income inequality, prioritizing efficiency over equity (Hoffman et al., 2020). This perspective argues that inequality is a natural by-product of market forces, and incentivizes effort, risk-taking, and innovation, which are the driving forces of Western economic success. It promotes policies that enhance efficiency – even if they result in unequal outcomes—as long as those who benefit could theoretically compensate those who lose. Liberals often dismiss concerns about the widening wealth gap as "the politics of envy", asserting that as long as inequality stems from fair market processes and equal opportunity, it is not a concern (Archer et al., 2022: 181). However, this stance (separating equity and efficiency) has been increasingly challenged, with many recognizing the potential negative consequences of excessive inequality on social stability and economic growth.

The liberal view contends that public policy should address inequality only when it threatens to provoke widespread resentment that could undermine the liberal order. It might now have reached that point in the United States. To understand the gradual, but staggering, growth in wealth and income inequality in the US, the following facts present a clear picture:

- In terms of wealth inequality, the real income of the bottom 50% of the population has grown only 20% since 1980, while the top 10% enjoyed 145% growth (McKay, 2023).
- As a primary measure of income inequality, the US Gini coefficient was 34.7 in 1980 and 41.5 in 2019 (inequality increased almost 20%) and 41.3 in 2022 (World Bank, 2024).
- The top 1% of earners in the US nearly doubled their earnings from 1979 to 2019, while the bottom 90%'s annual wages increased by a smaller amount (Gould & Kendra, 2021).
- Emphasizing the skewed income distribution (also see Fig. 2), America's richest 1% of households averaged 139 times as much pre-tax income as the bottom 20% in 2021, while the earnings of the top 0.1% income earners in the US increased by 345% from 1979 to 2019 (Congressional Budget Office, 2024).
- The COVID-19 pandemic has also deepened income inequality in the US. Low-wage workers, who faced higher job losses, reduced hours, and limited access to remote work, has been affected disproportionately. At the same time, wealthier individuals, particularly those in sectors like technology and finance, benefited from stock market gains and economic recovery policies, widening the wealth gap. Income and wealth inequality in the US is among the highest of any developed nation and continues to grow (Siripurapu, 2022).

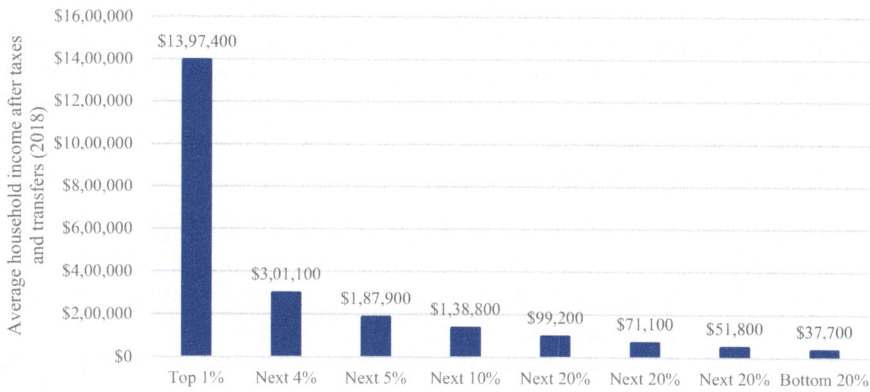

Fig. 2 Income distribution in the United States. *Source* Data from Congressional Budget Office (2024)

The US tax system has a redistributive effect, meaning that income inequality is lower when measured after taxes and transfers compared to pre-tax income, which actually mitigates the disparities in Fig. 2. The magnitude of redistribution in the US has been relatively limited compared to some other high-income countries. While labour income is the main driver of income inequality in the US, capital income has magnified the growth in inequality. The capital-to-labour income ratio has risen disproportionately among high-earning individuals. Stiglitz (2013: 124) therefore warns that, "When those at the bottom of the income distribution are at great risk of not living up to their potential, the economy pays a price not only with weaker demand today, but also with lower growth in the future".

Concentrated income and wealth lower overall economic demand, as wealthy households typically spend a smaller portion of their income compared to poorer households. This is even more pronounced in the case of the US and, additionally, the limited opportunities for low-income households have had a negative impact on economic growth and social mobility since 2020. Educational inequality, racial and ethnic disparities, geographic inequality, a stagnating minimum wage, outsourcing, and the fading power of labour unions have all contributed to low levels of economic mobility (i.e., moving up the economic ladder).

The rising wealth and income inequality in the US are generating more and more concerns as these factors lead to a higher cost of living for many, increased hardship, rising crime levels, mental illness, and social unrest (Johnston, 2024). On the last issue, significant economic disparities in the US have led to social fragmentation, increased polarization, and resentment among income groups, which weaken the sense of shared purpose and unity within American society. This was particularly evident in the run-up and during the 2024 US Elections. The political polarization has created a divided America, which is exacerbated by inequality, as different income groups have divergent interests. Inequality is further worsened by political factors, such as declining union density, changes in labour regulations, tax policy decisions (more tax cuts for high-income earners and corporations), and cutbacks in social welfare programs. Given his forceful approach, President Trump's new term of office could increase rifts in American society between the left and the right, which could increase domestic tension and raise fears of civil war. Further factors eroding social cohesion and fuelling political instability/division in the US—as these affect inequality—include:

- *Political influence*: income inequality has led to a higher concentration of political power among the wealthy, allowing them to shape policies in their favour, which creates the risk (and suspicion) of exacerbating inequality further and undermining democratic processes.
- *Technological change*: the emphasis on high-tech innovation have led to job losses in manufacturing and other industries, and have increased demand for highly skilled workers. In addition, the internationalization of financial markets, the rise of automation, and the growing pressure of remaining globally competitive are causing the US to not invest in reducing inequality to enhance productivity and innovation, thus falling behind other nations by not addressing this issue, affecting its economic and geopolitical influence.
- *Urban sprawl*: the uncontrolled and unplanned expansion of urban areas into surrounding rural land is closely linked to income inequality through its impact on housing, transportation, access to services, and economic mobility. This phenomenon in the US is characterized by scattered housing subdivisions, commercial strip malls, and wide stretches of roadways that often lack efficient public transportation and walkability. It presents significant social, environmental, and economic challenges, making it unsustainable. This pattern of development results in increased economic costs due to infrastructure expansion and maintenance, social segregation, and environmental damage from pollution and habitat loss. Furthermore, taking demographic factors like changes in household structures and marriage rates into consideration, as well as social issues (e.g., reduced access to urban amenities) and the fact that cities with higher levels of income inequality usually have higher rates of crime, the deepening of these disparities in the US presents serious risks.

A last major factor that arguably makes the US economy unsustainable relate to climate change and environmental challenges. The US, like other countries, is vulnerable to the impacts of climate change, such as extreme weather events and rising sea levels. Just in the past year (2024), four tropical cyclones, 17 severe storms (including tornados, baseball-size hail, and other abnormal-level storms), huge wildfires, droughts, severe floods, and a number of winter storm billion-dollar disaster events have affected the United States such that thousands have lost their lives and/or their livelihoods (Liao & Walls, 2024). In fact, the US has endured 11 billion-dollar disasters in 2024 and even more (a record number of) such weather and climate disasters in 2023. From 1995 to 2022, extreme weather events resulted in 13,857 fatalities, US$339 billion (adjusted to 2020 dollars) in property damage, and US$48.8 billion in crop losses. Between 1980 and 2024, the

US experienced 203 severe storms, 31 droughts, 66 tropical cyclones, 24 winter storms, 44 flooding events, 23 wildfires, and 9 freezes, each causing damages exceeding a billion dollars (adjusted for inflation) (NCEI, 2024). Just in May 2024, the National Weather Service's Storm Prediction Center received over 6100 reports of severe weather, including 475 tornadoes (NWS, 2024).

Apart from the damages, disruptions to supply chains, and the displacement of communities caused by these disasters that the US has to deal with constantly—challenges shared by many other countries—the larger issue is how this extends to long-term systemic risks that hinder growth, reduce productivity, and increase inequality (low-income communities are disproportionately affected by climate change due to limited resources for adaptation). It requires substantial (and low-return expectation) investment in climate adaptation, mitigation, and sustainable development to foster a resilient and equitable economy. Climate change is leading to more frequent and severe extreme weather events, imposing considerable costs on the economy.

For example, the estimated cost of damages from these events in the US has reached about $120 billion per year over the past five years (Vahlsing & Yagan, 2022). This cost is expected to rise significantly in the coming years, potentially reaching trillions of dollars annually in a few decades. Climate change is certainly making the US economy increasingly unsustainable by exacerbating environmental, social and economic vulnerabilities. The lack of sufficient investment in sustainable practices and infrastructure, which is not strongly prioritized by the Trump administration, could leave the US economy vulnerable to climate-related disruptions, such as extreme weather and resource scarcity.

The economic toll of climate change in the United States is not limited to immediate disasters, but also extends to long-term repercussions. It is projected to reduce the long-term growth rate of the US economy. Studies using both cross-country and within-country data for the US indicate that persistent increases in temperature above historical norms are associated with lower output growth (Bilal & Rossi-Hansberg, 2023; Mohaddes et al., 2023). It is estimated that a 0.01 °C annual increase in temperature above the historical norm could lower average per capita real Gross State Product (GSP)[1] growth by 0.0273 percentage points per year. Accumulated over time, this impact could translate into significant losses in GDP. It is projected to lead to a 1.0% reduction in the level of real GDP in 2050 due to climate change (Herrnstadt & Dinan, 2020). Inwardly, climate-induced instability places significant fiscal pressure on government budgets as federal and state governments are spending more and more on disaster relief, recovery, and infrastructure repair, plus related military expenses. Externally, the US economy is heavily reliant on global trade. Climate change impacts around the world can disrupt supply chains, increase transportation costs, and reduce demand for US exports, all of which can affect economic growth negatively.

---

[1] GSP growth measures the economic growth of a state (in the US) adjusted for inflation and population growth.

The negative impacts of climate change are felt across various sectors of the US economy. Agriculture, manufacturing, mining, construction, transportation, retail and wholesale trade, services, and government services are all susceptible to climate change impacts. For example, agriculture is negatively affected by extreme heat and excessive precipitation (leading to food shortages and higher prices), while manufacturing can experience costly production volatility due to climate variations. Extreme heat can reduce worker productivity, especially in outdoor industries (Bulut et al., 2023). In addition, climate change-related health issues, such as heat-related illnesses, respiratory problems, and the spread of diseases, can increase healthcare costs. Climate change can also lead to increased insurance costs and financial market instability. There are transition risks that stem from the shift to a low-carbon economy, which could result in the devaluation of assets tied to fossil fuels, potentially leading to financial instability.

The United States is experiencing a number of difficulties in the transition to carbon neutrality. It requires significant changes in both energy supply and consumer demand. On the supply side, increasing the share of renewable energy in the total energy mix necessitates rapid scaling up of production and deployment of renewable energy technologies, as well as expanding and upgrading electricity transmission infrastructure (Adua et al., 2021). On the demand side, rigid consumer habits in transportation, housing, and food consumption could limit behavioural plasticity and hinder the adoption of low-carbon practices.

Another difficulty is permitting and regulatory hurdles. There are complexities and delays associated with permitting processes for infrastructure projects, including those related to renewable energy and electricity transmission. Navigating multiple jurisdictions and satisfying a combination of environmental regulations, such as those under the National Environmental Policy Act (NEPA), take years and create opportunities for litigation from groups seeking to block projects. Other difficulties include economic dependence on fossil fuels, lack of political will and further political polarization, and the potential negative impact on US industrial competitiveness and global trade (e.g., as it implements policies to reduce carbon emissions, some polluting industries may relocate to countries with less stringent environmental regulations). The economic costs of transitioning to a low-carbon economy is extensive, especially given the size of the US economy. The following provides a summary:

- *Infrastructure investment costs*: investments in renewable energy infrastructure, such as wind, solar, geothermal power, and energy storage systems require substantial upfront costs; investments/upgrades to the national grid are needed to support distributed energy generation; expensive energy efficiency upgrades are necessary, relating to retrofitting buildings and industries to improve energy efficiency; and developing charging infrastructure for electric vehicles requires substantial investment. In 2023 financial year, the Biden administration allocated US$44.9 billion for climate action. In terms of R&D, significant funding is needed for innovation in clean energy technologies, carbon capture, and storage systems, as well as improvements in energy efficiency (Shao et al., 2021).

- *Job displacement*: Workers in carbon-intensive industries, such as coal mining and oil extraction, may lose jobs, requiring government programs for retraining to acquire new skills to adapt to the changing economy, and then redeployment into greener industries (Zabat et al., 2022). Such transition costs will then coincide with temporary disruptions in the economy, as capital shifts from traditional industries to renewable energy and sustainability-focused sectors affect GDP growth patterns and income generation.
- *Regulatory costs*: The implementation of stricter emissions standards, carbon taxes, or cap-and-trade systems (having to buy carbon permits to offset emissions) can increase compliance costs for US businesses, particularly those in manufacturing, transportation, and energy (e.g., upgrade their equipment and facilities to meet energy efficiency standards and eco-labelling certifications for their products).
- *Increased energy costs and supply chain disruptions*: The transition to clean energy often leads to short-term increases in energy costs. It can also disrupt supply chains as input costs increase and the cost of carbon-intensive production processes increase.
- *Stranded assets*: Transitioning away from fossil fuels may render coal, oil and gas reserves financially unviable, affecting energy-dependent industries and regions. Such a rapid shift could lead to the devaluation of assets, creating 'stranded assets' and possibly destabilizing financial institutions with exposure to these assets (Chen et al., 2022).
- *Consumer behaviour and social norms*: Americans' adoption of low-carbon practices and 'going green' is often slow due to concerns about aligning with certain social identities. Consumers tend to focus on the high immediate costs, discounting the long-term savings associated with low-carbon options like solar panels, weatherization, or heat pumps. The initial costs may in particular be a straining factor for lower-income households.
- *Inflationary pressures*: The cost of green investments and the shift in energy production may temporarily increase inflation due to higher production costs passed to consumers.

While few would argue against the need for adaption through comprehensive mitigation policies, including investments in renewable energy, energy efficiency, and carbon emissions reduction to address the root causes of climate change, the risks and costs these might have for the US economy suppress the political and corporate will power. The forces of competition and geopolitical tension further prevent drastic action. Achieving carbon neutrality might be a too challenging goal at the moment for the US. In truth, it is important to acknowledge the complexity of achieving carbon neutrality, because the transition will involve significant economic and societal changes.

Missed economic opportunities may be a concern, though, because slow adoption of renewable energy technologies risks the US falling behind in the global green economy, missing out on innovation and job creation opportunities. Other nations like China and India that are investing heavily in climate resilience and sustainable technologies may gain a competitive advantage. The US is therefore in a unique position, as it faces pressure from both sides (adaption and competition) to make real, substantial progress towards enhancing the long-term sustainability of its economy.

3 Towards De-dollarization?

Dollarization became a viable option as policymakers explored various monetary policies and exchange rate systems toward the end of the twentieth century. The widespread use of the US dollar has established it as the default global reserve currency. Since replacing the pound sterling in the 1920s as the reserve currency, the dollar is the single currency that dominated the global market in the past century. Now in the twenty-first century, the change in the global power balance is raising questions about the sustainability of the current global monetary order, anchored by the US, and the desire to create an alternative system that better reflect—and meet the needs—of the rising Eastern bloc of nations.

Trends in currency reserves actually indicate a gradual decline in the dominance of the US dollar, as emerging contenders begin to influence the future of the global financial system, albeit modestly. The share of US dollars in foreign exchange reserves fell from 73% in 2001 to 59% in 2021, and by 2023, dollar claims accounted for 54.59% of total foreign exchange reserves, well ahead of the euro at 20% and the yen at 5% (Bharat et al., 2024; IMF, 2024a). Hence, de-dollarization has been driven by concerns over the weaponization[2] of the dollar (US's use of economic sanctions), the GFC, and the Federal Reserve's response to COVID-19—all of which contributed to global distrust. This shift has been further accelerated by China's growing influence as a trading partner and investor. China and other BRICS+ countries are concerned about the dollar's stability and reliability, which is driving them to diversify their currency reserves.

The de-dollarization trend aims to develop alternatives to the existing system and promote reform. As the dollar's international role diminishes, we might already be seeing how the global financial system is evolving into a more multipolar structure. While it is unlikely that the dollar will soon be replaced as the reserve currency by another currency, the dollar's hegemonic status is not inherent, and cannot be taken as a given. The changing global financial landscape points to a possible change. A more balanced global financial system could potentially lead to greater economic independence for countries that are currently heavily reliant

[2] The term 'weaponization' refers to how a neutral but essential facility—the US dollar and SWIFT—have been turned into a weapon by one UN member state against another member state without the approval of the UN Security Council.

on the US dollar. Until now, underpinning world finance, the dollar controlled the global monetary cycle, subjecting non-US economies to US monetary policy (e.g., timing of interest rate adjustments). By accumulating reserves and engaging in foreign exchange interventions, countries can reduce their susceptibility to US monetary policy.

So, what is de-dollarization? It is the process by which countries and global institutions reduce their reliance on the US dollar for international trade, investment, and as a reserve currency (Liu & Papa, 2022). It involves diversifying away from the dollar in favour of other currencies, such as the renminbi, euro or even regional and alternative digital currencies or payment systems to settle transactions and hold reserves. By using more local currencies, it reflects a decline in the appeal and dominance of the dollar—or at least a decline in countries' willingness to use it as a reserve currency. De-dollarization furthermore involves establishing an alternative global financial and technological framework to achieve greater economic independence by bypassing reliance on Western-dominated systems, such as the SWIFT system for international trade and payments. The United States Department of the Treasury (USDT) is in charge of the SWIFT network, giving it significant influence over international financial transaction systems, including imposing sanctions on foreign individuals and organizations. It connects over 11,000 member banks in about 200 countries and territories globally. Currency worth trillions of dollars are sent across borders daily.

BRICS Pay is an example of an alternative system. It was launched in 2018 by the BRICS Business Council in a joint venture between financial and banking technology experts from the BRICS nations. It enables member states and other developing countries to make international payments more secure, transparent, and less expensive by doing transactions in their local currencies (Hilotin, 2024). BRICS Pay can help developing countries safeguard their economic sovereignty and expand economic and trade cooperation between member countries. It uses a blockchain-based platform designed to facilitate secure cross-border transactions within the alliance, which is aimed at increasing their bargaining power on the global stage. In fact, at the 2024 BRICS Summit in Kazan, Russia, President Vladimir Putin presented a mock-up of a supposed new BRICS currency and emphasised the bloc's commitment to exploring a common digital currency (Lee, 2024). He pointed out that BRICS members, particularly Russia and China, are driven to promote financial transactions using local and digital currencies to enhance economic independence and minimize exposure to Western financial systems—specifically to the US dollar in global trade. When implemented, the proposed BRICS digital currency is expected to operate just like a CBDC, being issued and regulated by each member nation's central bank. This would enable governments to oversee and manage the currency, avoiding challenges linked to decentralized cryptocurrencies.

This is exactly what President Trump warned about, even before starting his second term in office. Writing on 30 November 2024 on his social media platform, Truth Social, he threatened to impose 100% tariffs on countries that would create a new currency to rival or replace the US dollar. "We require a commitment from these countries that they will neither create a new Brics currency nor back any other currency to replace the mighty US dollar or they will face 100% tariffs and should expect to say goodbye to selling into the wonderful US economy," Trump said (Wearden, 2024). Reflecting his hard-line approach to trade and 'America First' policy, which aims to recalibrate global trade dynamics and reinforce US economic sovereignty, this comes off the back of his promise during his electoral campaign to impose harsh tariffs, like 60% on Chinese goods.

This is certainly raising the geopolitical and geoeconomic tensions as fears grow of the US dollar being used as a political weapon. It stems from the "geopolitical Triffin dilemma," named after economist Robert Triffin, who identified the tension between a country's need to run a current account deficit to provide liquidity to the global economy and its need to maintain domestic economic stability (Bharat et al., 2024). This geopolitical dilemma is an extension of the classic Triffin dilemma, which describes the inherent tension between the global role of a reserve currency and the domestic economic needs of the country issuing that currency. It considers the geopolitical implications of a country's dominant currency status: as a country's currency (like the US dollar) becomes more widely used globally, it can lead to increased geopolitical influence and power. This can, however, also lead to resentment and backlash from other countries, particularly if, in this case the US dollar, is used to impose economic sanctions or other punitive measures.

Hence, in view of future reserve utilization constraints, rather than a lack of fiscal room, it becomes the catalyst for a move away from dollar assets. The US needs to keep a delicate balance between economic power and geopolitical influence as it faces increasing pressure to maintain global financial stability while also addressing domestic economic challenges. Put differently, the balance also involves global stability vs. geopolitical rivalry: while preserving the dollar's global dominance to ensure economic and geopolitical influence, it also creates friction with rising powers and economic blocs (like China and BRICS+), potentially destabilizing the global financial system. The 'geopolitical Triffin dilemma' reflects the struggle between the US's need to maintain global dominance of the dollar for strategic advantage and the growing push from other countries to diversify away from it for economic independence and to reduce geopolitical risks. This tension is reshaping global economic governance and accelerate moves toward a multipolar financial system.

On the issue of the US weaponizing the dollar for strategic purposes, the USDT, in coordination with its Western allies, used the SWIFT system as a tool to isolate Russia economically following its full-scale invasion of Ukraine in February 2022. Here is how it was leveraged:

- Numerous Russian and Belarusian banks were disconnected from the SWIFT network as part of economic sanctions imposed by the US and its allies.

- This removal severely limited Russia's ability to conduct international transactions, making it difficult for them to engage in global trade and finance (e.g., having access to foreign capital).
- It disrupted Russia's financial connectivity with global markets, affecting everything from corporate operations to international trade settlements.
- The sanctions specifically targeted key Russian financial institutions, such as Sberbank and VTB, which play a critical role in the Russian economy—restricting their ability to process cross-border payments and access foreign currencies.
- Blocking Russian banks from SWIFT made it difficult for Russian companies to export goods and services or import critical items—effectively isolating Russia from a significant portion of the global financial system.
- Alongside SWIFT disconnections, the USDT imposed asset freezes and targeted financial sanctions on Russian oligarchs, government officials, businesses, and entities, creating additional hurdles for Russia to bypass the SWIFT-related sanctions.
- Using SWIFT as a sanctioning tool was part of a broader, coordinated effort by the US, the EU, the UK, Canada, and other allies to exert economic pressure on Russia.
- The US and its allies also used other sanctions to weaken Russia's ability to finance its military operations; such as, export controls (banning high-tech goods and dual-use technologies that could be used for military purposes); travel bans; and trade restrictions (imposing tariffs and quotas on Russian goods, particularly energy exports).

By limiting Russia's access to the global financial system, the US aimed to deter further aggression and weaken Russia's economic and political influence. These sanctions underscore the role of financial systems in modern geopolitics and highlights the leverage Western nations have over the global financial infrastructure. However, they also incentivize Russia and its allies to explore de-dollarization strategies and strengthen regional economic partnerships. For instance, Russia sought to mitigate the impact by expanding the use of alternative payment systems, such as its own SPFS (System for Transfer of Financial Messages), which serves 400 users (Mayer, 2024). It also made use of Chinese financial networks like CIPS (Cross-Border Interbank Payment System), owned by The People's Bank of China (reaching more than 3000 banks across 167 countries and regions). In this way, de-dollarization is seen in the East as a counter-strategy to Western/US aggression and the weaponization the dollar. It shows the dangers of using the dollar-based financial system for geopolitical leverage. Such economic statecraft is becoming a dangerous strategy in today's more balanced multipolar global order. Every geopolitical decision has larger consequences.

Apart from promoting local currency transactions and developing alternative payment systems, the BRICS+ countries are also exploring other strategies to reduce their reliance on the US dollar. As referred to earlier, the creation of new regional financial institutions like the New Development Bank (NDB) is part of their de-dollarization strategy to provide alternative financing options for member countries and other emerging economies in their own currencies to reduce exchange rate risks and reliance on the dollar (Kumar et al, 2024). The inclusion of new members, such as Egypt, the UAE, and Iran, enhances the NDB's capacity to promote non-dollar transactions in a broader region.

This initiative challenges the traditional control of Western institutions over development finance and promotes greater financial autonomy for BRICS+ members. These countries have, in terms of trade, implemented agreements to settle bilateral and multilateral trade transactions in their local currencies rather than the US dollar. For example, India and Russia are using the Indian rupee and Russian rouble for defence and energy trade (oil and gas exports), while China is increasingly using the yuan/renminbi in trade with Brazil and other BRICS+ countries (Shah et al., 2024). In fact, China is actively promoting the yuan as a global trade currency by expanding its use in BRI projects and signing currency swap agreements with other countries. By allowing trade to bypass the dollar through such agreements, Russia, Iran and some Southeast Asian countries are increasingly settling trade in local currencies or yuan, reflecting Asia's efforts to insulate against Western financial sanctions. In this way, the BRICS+ countries foster closer economic ties with other emerging economies, creating alternative trade and investment networks. China is even testing its digital yuan (e-CNY) in cross-border transactions relating to joint infrastructure projects and development initiatives. BRICS+ countries are investing in alternative trade routes and infrastructure projects to reduce dependence on dollar-dominated supply chains (Caldararo, 2023). BRICS central banks are reducing the share of US dollars in their foreign exchange reserves and increasing holdings of other currencies (yuan, euro) and assets like gold. China and Russia have been accruing gold reserves as a hedge against dollar volatility and are exploring the use of gold as a means of exchange and store of value—even using gold to price and settle commodity trade.

BRICS+ countries are leveraging a multi-pronged strategy to reduce dollar dominance in trade, finance, and investment. While these efforts have gained momentum, achieving significant de-dollarization will depend on overcoming challenges like limited liquidity of alternative currencies, geopolitical rivalries, and the entrenched dominance of the US dollar in global markets. What is adding to the momentum is other emerging economies also following suit. Since 2023, Argentina has joined Brazil in paying for Chinese imports using the yuan instead of US dollars in an effort to safeguard its diminishing reserves of dollars. After China and Argentina declared in April 2023 that their currency swap line had been opened, it enabled Argentina to use the equivalent of 1.04 billion yuan to settle payments for Chinese imports in May (Carmody, 2024).

Following Sergio Massa's visit to China, the swap line was expanded in June to the value of US$18 billion over the next three years. After Russian President

Putin signed a decree on 1 April 2022, mandating unfriendly countries to pay for natural gas imports in roubles, four European gas companies still made trade payment settlements in roubles, despite European leaders having rejected paying for deliveries in roubles because it would undermine the sanctions they imposed. In 2013, Australia made an agreement with China to trade in national currencies (Siddiqui, 2024). In August 2018, Venezuela announced its decision to price its oil in roubles, yuan, euros, and other currencies. In December 2022, Mauritius and Sri Lanka started using the Indian rupee for international trade. One of the main reasons for less developed countries to de-dollarize is that most commodities are priced in dollars, meaning they are importing inflation that would otherwise accrue to the US.

There are significant complexities and challenges associated with establishing a truly alternative monetary system. While progress has been made in developing institutions and mechanisms, the dollar's deep liquidity, global acceptance, and the strength of its supporting institutions remain significant hurdles. Differing economic strengths and priorities among BRICS members also present challenges in coordinating a unified de-dollarization strategy. For example, countries like Brazil, South Africa and India, with closer ties to the US, may be less enthusiastic about a complete break from the dollar-based system. Notably, the de-dollarization efforts are not solely about replacing the dollar, but also about challenging US hegemony in the global financial system. The BRICS countries therefore keep making concerted efforts towards shaping an alternative monetary system, seeking a more multipolar world order where power is more evenly distributed and their interests are better represented. Additionally, given the declining share of the US in global output, the dollar losing market share to other currencies, and the rise of new financial expertise like blockchain technology, factors that would have sustained the dollar's dominance are in decline.

An important question is whether de-dollarization is inevitable. This is a matter of intense debate, but most agree that it will not be the case in the short term. There will not be a sudden and complete replacement of the US dollar in the near future, because its dominance is deeply ingrained in the current global financial system. Hence, the question of inevitability has to do with what happens to the dollar during the gradual shift towards a more multipolar system, which could take decades. However, a sudden decline in demand for the US dollar could have devastating effects on Americans, potentially sparking a dollar crisis that could result in severe inflation or even hyperinflation (Giustra, 2023). This could trigger a cycle of escalating debt and excessive money printing. This is arguably the worst-case scenario and could speed up de-dollarization. More realistically, perhaps, is a gradual move in this direction, driven by the following factors:

- The weaponization of the dollar, which is eroding trust and pushing countries like Russia, China, and BRICS+ members to reduce their reliance on it.
- The trend towards a multipolar world, with emerging economies challenging the United States' dominance in international trade, finance and in global institutions.

- The rise of alternative currencies like the Chinese renminbi, the Indian rupee and the Russian rouble and their growing role in international trade and finance, along with the development of regional payment systems.
- The rise of new financial technologies that now provide the infrastructure for a shift away from the dollar-based system, are reducing transaction costs, enhancing transparency, and making it easier for countries to use alternative currencies for cross-border payments.
- The US dollar's dominance is not a historical constant, as was seen with the British pound sterling. Shifts in global power and currency regimes are essentially inevitable.

Another question is how de-dollarization will change the global monetary and financial order. A reduction in the dominance of the US dollar in international trade, finance and reserves has the potential to reshape the global monetary system significantly. The first consequence would be the diversification of reserve currencies as central banks worldwide start increasing their holdings of alternative currencies such as the euro, yuan, or emerging regional currencies. Secondly, the emergence of a multipolar currency system where countries will prioritize using local or regional currencies for trade agreements, creating multiple monetary zones instead of one dollar-centric system, resulting in greater reliance on regional trade blocs. It may also present challenges, as managing a multipolar system requires greater coordination among nations and international institutions. Thirdly, a shift in financial centres and influence, resulting in a redistribution of global financial power alongside New York and London, shaping new borrowing and lending patterns. This could impact global financial regulations, infrastructure, and the concentration of economic power, even leading to a more decentralized financial system (Falarti & Naqvi, 2024). Fourthly, there will be a geopolitical realignment and new alliances, which could lead to a more fragmented world with competing currency blocs, currency wars and increased geopolitical tensions, particularly between the US and emerging economies challenging its dominance. De-dollarization poses a direct challenge to the US's global influence, reducing its economic and political leverage.

Fifthly, it will have a significant impact on global financial stability, which is open to debate. Reducing reliance on a single currency could enhance stability by diversifying risks and mitigating the impact of shocks originating from the US. On the other hand, a transition to a multipolar system could create volatility and uncertainty, potentially leading to financial instability, especially if not managed effectively. Sixth, it might present new opportunities for emerging economies to enhance their economic autonomy and global influence (Thiagarajan et al., 2023). By reducing their reliance on the dollar, these countries can mitigate their vulnerability to US economic policies and potentially gain greater control over their monetary and financial systems. This could also create opportunities for regional integration and the development of alternative financial institutions led by emerging economies. Seventh, the role of technological innovation will become much more important. Technological advancements, particularly in digital currencies and payment systems that is more efficient, secure, and widely accessible is crucial for reducing reliance on traditional dollar-based infrastructure and enabling the broader use of alternative currencies (Caldararo, 2023). CBDCs and blockchain-based platforms are key areas of innovation, with the potential to reshape cross-border payments and challenge the dollar's dominance.

What risks do de-dollarization pose to the US economy? It is clear that the impact would be predominantly negative. A reduced global role for the dollar could lead to higher borrowing costs for the US government, a diminished ability to run persistent trade deficits, and less influence over international financial regulations. A decline in the dollar's role could reduce the effectiveness of US sanctions and limit its ability to shape global financial policies (Nikolic et al., 2023). This could necessitate a reassessment of US foreign policy strategies and a potential shift towards greater multilateralism to maintain its global standing. Given the US's current high debt levels, it cannot afford an increase in borrowing costs. As countries reduce their reliance on the dollar, demand for US Treasury bonds, a traditional safe-haven asset, could decline. This reduced demand would force the US government to offer higher interest rates to attract investors, making it more expensive to finance its debt.

Additionally, if central banks decide to sell off their holdings of US Treasury bonds to strengthen their own currencies, it could further reduce bond prices and drive up US interest rates. This scenario would create a vicious cycle of higher borrowing costs, increasing debt levels, and significant strain on the US economy (Siddiqui, 2024). De-dollarization could also lead to a weakening of the US dollar in foreign exchange markets. As demand for dollars declines, its value relative to other currencies could depreciate. This devaluation would make imported goods more expensive for American consumers, potentially leading to inflation and reduced purchasing power. While a weaker dollar could make US exports more competitive in the short term, the negative consequences for consumers and the broader economy could outweigh these benefits.

In case of de-dollarization, there would be an erosion of the unique benefits ('exorbitant privilege') that the US enjoys due to the dollar's status as the world's reserve currency. These benefits include the ability to borrow cheaply in global markets, create money inexpensively, and using the dollar's global influence for its own economic advantage. De-dollarization could also negatively affect US financial institutions and markets. As global financial activity shifts away from the dollar, US banks and financial institutions may see a decline in their business and profitability (Liu & Papa, 2022). Reduced demand for US Treasury bonds will create challenges for US commercial banks holding significant amounts of these assets.

In addition, a less dollar-centric global financial system could diminish the influence of US financial markets and potentially lead to a loss of jobs and economic activity in the US financial sector. De-dollarization would force the US to adapt to a more competitive global economic environment. As alternative currencies and payment systems gain traction, the US would need to innovate and enhance its competitiveness to maintain its economic standing. This could involve promoting greater efficiency in its financial system, investing in new technologies, and potentially reconsidering its approach to global economic governance to better accommodate the interests of emerging economies. It is the US's response to de-dollarization, the development of viable alternative currencies, and the evolving geopolitical landscape that will determine the extent to which the risks will materialize and impact the US.

Another key question is how the petrodollar system will be affected by de-dollarization. The term 'petrodollar' refers to US dollars earned by oil-exporting countries through the sale of oil, which is typically priced and traded in US dollars. Oil is the most widely traded commodity in the world, and by pricing it in dollars, there is a guaranteed global demand for dollars, which protects its value. This system originated in the 1970s following the collapse of the Bretton Woods system, where the US dollar was pegged to gold, and has since been a cornerstone of US economic and geopolitical dominance. Under the petrodollar system, oil-producing countries accumulate large reserves of US dollars, which they then reinvest in US Treasury bonds and other dollar-denominated assets (Bharat et al., 2024). This system is one of the main pillars of the dollar's status as the global reserve currency.

De-dollarization would give momentum to BRICS+ and OPEC countries' efforts to challenge US dollar dominance. If these countries would shift away from pricing oil in US dollars, it will undermine the petrodollar system and accelerate the adoption of alternative currencies like the Chinese yuan and the Russian rouble for oil trade. This would erode the dollar's dominance in global energy markets. Major oil producers like Saudi Arabia, Russia, or Iran could expand bilateral trade agreements that bypass the dollar, reducing its centrality in oil markets.

As oil trade shifts towards other currencies it could bolster their international standing. The de-dollarization of the oil market would have serious disruptive consequences with ripple effects throughout the global economy. The reduced demand for dollars in global markets would drastically weaken the dollar's value, leading to higher import costs and inflation, and reduce its role as the global reserve currency. If oil transactions no longer require dollars, countries would need fewer dollar reserves, which would further decrease global demand for US currency. It will also weaken US influence and challenge its ability to exert economic pressure through dollar-based sanctions. This will cause a shift in geopolitical alliances as oil-producing nations may align more closely with countries promoting alternative trade currencies, such as China and Russia.

It will significantly reduce the US's influence in global energy policy. Foreign investment in US Treasury securities will also decline, leading to higher borrowing costs for the US government. The absence of a single currency for oil trade could lead to fragmented pricing mechanisms, increasing volatility in oil prices and complicating global market stability. Different regions may develop their own benchmarks for oil pricing, which would reduce the dominance of US-linked standards like Brent Crude and WTI (West Texas Intermediate). While de-dollarization is a complex and long-term process facing considerable obstacles, the efforts of

BRICS + and OPEC countries to reduce reliance on the US dollar represent a substantial challenge to the existing global financial order.

The likely outcome is not the end of the dollar's global role, but a transition toward a multipolar currency system where the dollar coexists with other major currencies like the yuan and euro. This shift may take decades, driven by the interplay of economic, political, and technological factors. BRICS+ nations should carefully evaluate how the United States, particularly under Donald Trump's presidency, might respond to abrupt moves away from the dollar. History shows that shifts in global economic power rarely occur without significant conflict or warfare. However, despite resistance from the US, de-dollarization will continue and it will need to make adjustments, as much of the non-Western world seeks a more equitable trading system, free from the vulnerabilities of dollar weaponization and dominance. It is no longer a matter of if, but when.

4 Is BRICS+ A Challenge to US Hegemony?

Before answering the intriguing question above, let's first consider what BRICS+ exactly is. It can be described as an informal grouping or bloc of countries that formed a partnership and has developed into an intergovernmental organization that aims to increase economic and geopolitical cooperation among its member countries and increase their economic and political standing in the world (Ng, 2024). Having started to expand beyond the five initial countries, BRICS+ is a powerful grouping of the world's leading emerging market economies. There is no indication that these countries would become a political alliance like the EU or even a formal trading association, but they do share a number of joint priorities to achieve their aim. For this, their governments meet annually at formal summits and have been coordinating multilateral policies since 2009. BRICS relations are primarily guided by principles of non-interference, equality, and mutual benefit. An important feature that they also share is being a counterbalance to the West. Working together, they have implemented competing initiatives, such as the NDB, the BRICS Contingent Reserve Arrangement (CRA), BRICS Pay, the BRICS Joint Statistical Publication, and the BRICS basket reserve currency through which they are seeking to reshape global economic and political norms.

On closer investigation, while still an evolving grouping of nations, of which some are authoritarian, there are three possible interpretations of the purpose of BRICS+. According to Van Nieuwkerk (2024), it can be one of three options: (1) a club of self-interested members—a sort of Global South cooperative functioning as a self-help organisation; (2) a reforming bloc with a more ambitious goal of improving the workings of the current global order; and (3) a disrupter, preparing to replace the Western-dominated liberal world order. Examining the commitments made during the 2024 meeting in Kazan, Russia, it can be argued that BRICS+ positions itself primarily as a pragmatic reformer with global reordering ambitions. It reflects the perspective of Global South leaders regarding the current global order and their aspirations to shape a new one. This shift comes as the world transitions away from the financially dominant yet morally waning Western-led order, spearheaded by the US, toward a multipolar system where the East assumes a central role.

On the question of whether BRICS+ is or could be a challenge to US hegemony, there appears to be significant substance to observations that it actually could be—if not now, then later. With the group currently dominated by China, containing about 70% of the organization's total GDP, and China's growing collaboration with Russia and oil-exporting Arab states, new hegemonic aspirations, anchored in the East, seems to be quite strong. There are a number of indications in this respect—not in the sense of seeking to overthrow the existing global order—but rather a 'within-system' challenge, aiming to reform the current system to be more inclusive and multipolar. Three broad categories can be identified: economic challenges; geopolitical challenges; and strategic and technological challenges. In terms of economic challenges, the first is the de-dollarization referred to, in which the BRICS+ countries are actively pursuing alternatives to the US dollar in trade and finance, reducing their reliance on it and promoting the use of local currencies in trade and investment, which undermines the dollar's dominance and challenges the US's economic leverage. They remain steadfast in their efforts to advance a more diversified currency system, based on an alternative payment system and the use of digital currencies.

Secondly, BRICS+ is a rapidly emerging economic bloc. Together, these nations account for almost 50% of the world's population and over 30% of global GDP (in PPP terms), creating a significant counterweight to Western economic institutions. Due to their rapid economic growth, BRICS's share of global goods exports grew from 10.7% in 2000 to 23.3% in 2023, while they produced 30% of the world's oil by 2023 (including that of the new members) (IMF, 2024b). Accounting for a third of global trade, BRICS+ is almost surpassing the G7. Such economic power translates into political influence, giving them more leverage to demand changes in how the global economy is governed. Their newly created institutions such as the NDB and AIIB, as well as pushing for more voting rights and representation in the IMF, World Bank and WTO, are testament to their efforts to create alternatives and a completely reformed global economic governance architecture, as they keep fostering South-South cooperation.

Greater representation and decision-making power in international institutions—and less dominance by the US and neoliberal policies—are non-negotiables for them. Offering development assistance and partnerships based on non-interference and no conditionality principles, in contrast to traditional Western aid practices, gives the BRICS+ countries stronger bargaining power in promoting South-South cooperation. With BRICS+ nations integrated into the existing global capitalist system, and them benefiting from and contributing to its stability, they are indispensable to any forward planning for the global economy. By building stronger economic ties among themselves and with other developing regions, BRICS+ nations are actively reducing US influence in global trade flows. China's BRI, in particular, and growing trade relations with Africa, Latin America, and

Southeast Asia are realigning investment and trade flows, which is a significant challenge to US economic dominance.

The geopolitical challenges that the BRICS+ nations pose to the United States are firstly: on all fronts, BRICS+ is promoting a multipolar world that challenges the unipolar dominance of the US. BRICS+ offers an alternative to US-led alliances like NATO (the North Atlantic Treaty Organization) and the G7, giving countries that are disillusioned with Western policies a platform to voice their concerns and align on shared goals. New market access and investment opportunities play a vital part in swaying interest. With resource-rich members like Russia and Brazil, BRICS can influence global energy markets, potentially reducing US control over critical resources like oil and gas, particularly through partnerships with OPEC nations and moves away from the petrodollar system. Secondly, the growing political and diplomatic influence of BRICS+ nations, specially China, is challenging US leadership in various regions of the world. One specific example of this is in the MENA (Middle East and North Africa) region. BRICS engagement in this region has disrupted traditional alliances and created new geopolitical dynamics.

For instance, Russia's military intervention in Syria and its strengthened ties with countries like Iran and Turkey represent a direct challenge to US influence in the region (Özekin & Sune, 2023). BRICS's influence has emboldened regional actors like Iran and Turkey to pursue more independent foreign policies, often at the expense of Western interests. These regional powers leverage their relationships with BRICS nations to counterbalance US pressure. BRICS's state-led development model presents an alternative to the neoliberal policies promoted by the US and its allies. This model, emphasizing state intervention and control over strategic sectors, resonates with countries seeking greater economic autonomy and control over their development trajectories. BRICS+ nations have helped countries like Iran and Russia circumvent US and Western sanctions through trade and investment partnerships. This undermines US efforts to isolate and pressure countries that do not align with its foreign policy objectives.

In terms of strategic and technological challenges posed to the United States, BRICS+ members, particularly China and India, are investing heavily in the development of independent technological ecosystems to reduce dependence on Western technology (Türker, 2024). This includes advancements in telecommunications (e.g., China's Huawei), renewable energy, and digital currencies. China and India's rapid advancements in areas such as AI, 5G, quantum computing, and space technology pose a direct challenge to US technological supremacy.

This competition has serious implications for military capabilities, economic competitiveness, and the future of global innovation. China's efforts to build alternative digital infrastructure through initiatives like the BRI, including undersea cables and data centres, could significantly undermine US control over critical information flows (Caria, 2022). Concerns about cybersecurity threats and espionage activities originating from some BRICS+ nations, specifically China and Russia, are troubling to the US. These activities raise anxieties about the security of US critical infrastructure and intellectual property, adding another dimension to the strategic competition between the US and BRICS. By expanding into areas like MENA, BRICS+ nations are countering strategic US and European interests by making their escalating technology infrastructure easily accessible.

Although the rise of the BRICS+ nations represents a shift in the global balance of power and challenges the traditional US-centric world order, at the moment BRICS+ does not represent a fully formed counter-hegemonic bloc. While it exhibits characteristics of a counter-hegemonic force, it lacks the cohesion, shared ideology, and capacity to provide global public goods necessary to establish a truly alternative system. There are a number of critical limitations that hinder BRICS+ from being a unified or immediate threat to US hegemony, which include:

- *Lack of a unified vision and identity*: The heterogeneity within BRICS+, marked by diverse political systems, economic structures, and foreign policy orientations of its member states, prevents it from having a 'BRICS Consensus' (as opposed to a Washington Consensus). This diversity hinders the development of a cohesive group identity and a shared vision for a new global order (Nwaorgu & Nwankwo, 2015). For instance, Russia and China have often aligned against Western-backed UN resolutions, while other members like Brazil and India have adopted more balanced approaches, securing their Western interests.
- *Internal divisions and tensions*: The recent expansion of BRICS further exacerbates the challenge of maintaining cohesion (Farooq et al., 2020). There are existing tensions and historical rivalries between members as potential obstacles to effective collective action. Examples include the border clashes between India and China as well as the rebel invasion in Syria at the end of 2024, where Iran and Russia backed Assad's forces while Turkey supported the invading rebel groups. The differing stances on BRICS expansion, with China pushing for a larger membership while others prioritize a more robust stance on UN reform, also illustrate internal divisions.
- *Inability to provide global public goods*: A key characteristic of a hegemonic power is its ability to provide global public goods like security, a stable currency, and a functioning trading system (Öniş et al., 2020). BRICS+ currently lacks this capacity and the formal mechanisms and resources to effectively implement collective action and enforce agreements. For example, while initiatives like the NDB and CRA offer alternatives to Western-dominated financial institutions, they are not yet on par with the scale and influence of the IMF or World Bank.
- *Other challenges and inconsistencies*: Competition for resources and influence as well as conflicting interests often create tensions among BRICS+ members. BRICS's influence rests primarily on its economic clout and geopolitical manoeuvring (Firdaus et al., 2023). It lacks the soft power resources, such as cultural influence and widespread appeal of its norms and values, to effectively challenge the US-led order on an ideological level. Despite its critique of the neoliberal world order, BRICS remains deeply integrated into the existing

global capitalist system. This dependence limits its ability to challenge the system fundamentally and creates contradictions in its actions. BRICS's economic success is largely due to its participation in global trade and investment flows, often on terms set by Western-dominated institutions. This creates a contradiction between its desire for greater autonomy and its reliance on the very system it seeks to reform.

5 Conclusion

Increased geopolitical tension between the US and BRICS+ has significant implications for global sustainability across economic, social, and environmental dimensions. Such geopolitical rivalry can hinder international cooperation on crucial sustainability issues, dividing the nations between East and West. For example, disagreements over climate action, environmental regulations, and resource management can be exacerbated. In terms of economic sustainability, the fragmentation of global trade is a major concern, where rising tensions may lead to trade blocs and reduced multilateral cooperation, impacting global supply chains and economic efficiency (Ng, 2024). Short-term national interests may overshadow long-term sustainability goals. Currency volatility would increase as BRICS+ efforts to promote alternatives to the dollar could destabilize the global monetary system, leading to exchange rate uncertainty and financial instability. Development challenges could arise as countries caught in the crossfire may struggle to secure investments or favourable trade terms, particularly in developing regions dependent on both the US and BRICS+. In addition, escalating tensions can divert resources away from addressing critical sustainability challenges like climate change and poverty. Increased military spending, for example, diverts funds from investments in renewable energy, sustainable infrastructure, and social programs.

With regards to social sustainability, inequality is expected to increase in not only the US and BRICS+, but also in the rest of the world. A fractured global economic order could exacerbate wealth and opportunity disparities between nations, leaving smaller economies marginalized. Geopolitical rivalries may intensify internal divisions within countries, affecting governance (increased political instability), public welfare, and social cohesion. Escalating tensions could furthermore increase the risk of military conflict, which would have devastating humanitarian and environmental consequences. Regarding the impact of growing geopolitical tension between the US and BRICS+ on environmental sustainability, rivalries and competition for resources could escalate the exploitation of natural resources, particularly in regions with strategic importance, leading to environmental degradation (Türker, 2024). Climate cooperation on global issues, such as addressing climate change, could diminish as the US and BRICS+ nations focus on advancing their divergent geopolitical agendas over pursuing collective environmental goals. Achieving the SDGs would be much more problematic in a global climate of increasing geopolitical tension.

Growing geopolitical tensions between the US and BRICS risk undermining the cooperative frameworks needed to address global sustainability challenges. That is why it is crucial for the US and BRICS+ nations to find ways to manage their differences and cooperate on global challenges, such as climate change and sustainable development. The US will need to adapt its strategies and policies to navigate this evolving global landscape effectively. Fostering dialogue and maintaining multilateralism remain essential for ensuring a stable and inclusive global order. Open and transparent communication channels between the US and BRICS+, particularly with China, are crucial for managing differences and preventing misunderstandings. Dialogue should focus on identifying and ring-fencing common interests, such as climate change, global health security, and non-proliferation, where cooperation is possible despite broader strategic competition. Both the US and BRICS+ should adhere to the principles of national sovereignty and non-interference in the internal affairs of other states. This includes refraining from actions that undermine the territorial integrity or political independence of other nations.

For sustainable development, the US and BRICS+ should engage in healthy competition in the technological realm, while establishing clear rules of the road to prevent a destabilizing arms race or technology war. This includes promoting open and fair access to global markets, protecting intellectual property rights, and cooperating on international standards for emerging technologies. Dialogue on cybersecurity and responsible use of AI can help mitigate risks and build trust. Preventing an escalation of geopolitical tension between the US and BRICS+ will require a shift in mindset on both sides. The US needs to embrace the reality of a multipolar world and pursue a more collaborative approach to global leadership. BRICS+ nations need to demonstrate a commitment to responsible global citizenship and avoid actions that are perceived as threatening or destabilizing. Building a more stable and cooperative relationship will require both sides to find common ground, manage differences peacefully, and focus on shared challenges facing humanity.

Drastic interventions are needed by the United States to make its economy more sustainable—not just for the sake of the US, but also for the rest of the world. The new Trump administration seems much focussed on addressing some of the most chronic US economic problems, which include several sustainability-related interventions. President Trump has even created a new Department of Government Efficiency (DOGE), of which billionaire Elon Musk took up the lead, to cut back mercilessly on government spending in order to address the US debt crisis and its gigantic fiscal deficit. How successful it will be remains to be seen. To make its economy more sustainable, the US would need to implement a series of transformative steps across fiscal, environmental, and social domains, such as:

- *Reduce US national debt*: Implement comprehensive fiscal reforms to decrease the national debt through a combination of spending cuts and revenue increases. Prioritize investments in infrastructure, education, and healthcare over excessive military spending. Prioritize spending on essential services and reduce spending on non-essential programs. Increase taxes on the wealthy and corporations to reduce the national debt.

- *Promoting sustainable consumption and production patterns*: The US needs to encourage a shift towards more sustainable consumption and production patterns that reduce both waste and resource depletion. This could involve promoting energy efficiency, reducing reliance on single-use plastics, and supporting circular economy models that prioritize reuse and recycling. This would require a combination of policy incentives, consumer education, and technological innovations.
- *Comprehensive climate action:* Implement a robust carbon pricing mechanism, such as a carbon tax or a cap-and-trade system, to significantly reduce greenhouse gas emissions and incentivize green innovation. Accelerate the transition to renewable energy sources by phasing out fossil fuel subsidies and investing heavily in renewable energy infrastructure and clean technology. Upgrade energy efficiency standards for buildings, vehicles, and industries. Initiate a Green New Deal: implement a large-scale investment program in green technologies, energy efficiency, and sustainable infrastructure, creating new jobs and reducing reliance on fossil fuels. The US needs to invest in climate resilience, i.e., in infrastructure and technologies that can withstand the impacts of climate change, such as sea-level rise, extreme weather events, and water scarcity. This would involve strengthening coastal defences, upgrading infrastructure to withstand extreme temperatures and storms, and developing drought-resistant crops.
- *Revitalize domestic manufacturing and prioritize inclusive growth*: Promote reshoring of critical industries to reduce dependency on foreign supply chains and bolster economic resilience. Provide incentives for sustainable manufacturing practices, such as producing solar panels, electric vehicles, and wind turbines, which can generate new jobs in clean energy manufacturing and related supply chains. Enhanced focus on high-tech manufacturing, such as semiconductors, green energy technology, and advanced machinery will also create high-paying jobs in specialized fields. Domestic manufacturing stimulates demand for raw materials, components, and intermediate goods, creating more jobs in industries such as mining, chemicals, and parts production. Manufacturing has a strong economic multiplier effect, meaning jobs created in this sector indirectly lead to job creation in other parts of the economy, such as retail, healthcare, and education, as workers spend their incomes locally. For inclusive growth, develop policies that address income inequality, such as progressive tax reform, expanded social safety nets, affordable housing initiatives, and stronger labour unions. In addition, increase access to education and skill development to enhance workforce competitiveness.
- *Enhance trade policy and reform financial systems*: Re-evaluate and modernize trade agreements to include provisions for sustainability and labour rights. Diversify trading partners to reduce economic risks tied to geopolitical tensions. Regulate financial markets to reduce speculative bubbles and promote long-term

investment. Encourage greater savings by offering incentives for retirement and education accounts.

- *Foster technological innovation*: Support research and development in emerging sectors like AI, biotechnology, and green energy (e.g., solar, wind, and geothermal power). Advancements in energy storage technologies like batteries are crucial for the widespread adoption of renewable energy sources. Improve digital infrastructure to support the growth of the digital economy. Technological advancements can improve recycling technologies and develop more sustainable waste management solutions. Technologies like drones, sensors, and AI can optimize sustainable agricultural practices, reducing water usage, minimizing fertilizer runoff, and increasing yields.
- *Reinforce democratic institutions and strengthen global leadership*: Ensure transparent and accountable governance to build public trust and attract sustainable investments. Combat corruption and strengthen the rule of law to create a fair economic environment. Engage in multilateral efforts to create a more balanced global economic system. Lead initiatives that promote global sustainability, including carbon neutrality, biodiversity preservation, circular economy practices, the Global Compact,[3] and attaining the SDGs.

Whether the United States will implement all the necessary changes to make its economy more sustainable remains to be seen, but one thing is for sure: unless it is done quite soon, the US will be overtaken at some point by the rising giants of the East. Countries in the East are increasingly recognizing the strategic importance of economic sustainability and are taking steps to integrate environmental and social considerations into their development strategies. Their efforts to balance rapid growth with long-term resilience is constantly improving. China, India, and other Eastern economies are actively embedding economic sustainability into their growth strategies. However, the degree of prioritization and success depends on navigating development challenges, geopolitical factors, and global cooperation. There is a strong emphasis on green development, but because most Eastern economies, especially China and India, are still undergoing rapid industrialization, it presents significant environmental challenges. Ensuring energy security while transitioning to a low-carbon economy is a major concern for these countries. China has made significant strides in renewable energy, becoming a global leader in solar and wind power. India is making strong progress on sustainable agricultural practices, such as organic farming and water conservation, to improve food security and reduce environmental impact. Apart from dealing with high pollution levels, many Eastern nations are still addressing poverty and infrastructural deficits, which can clash with sustainability goals. The latter does, however, remain

[3] The Global Compact is a UN-initiative that encourages businesses worldwide to adopt sustainable and socially responsible policies. It focuses on human rights, labor standards, environmental protection, and anti-corruption.

a major focus for Eastern countries as they work to ensure that the benefits of economic growth are shared equitably and that environmental policies do not impact vulnerable populations disproportionately.

On the issue of de-dollarization, while it is not foreseeable in the short to medium term, the trends toward a more multipolar financial system are undeniable. Factors like geopolitical rivalries, alternative currency (and payment) systems, and technological advances may gradually weaken the dollar's dominance, but replacing it entirely requires overcoming immense structural and trust-related barriers. Economic instability (e.g., high inflation in Russia and India), internal tensions (e.g., China-India disputes), and geopolitical risks restrain rapid BRICS progress, making stable macroeconomic and political policies vital. Yet, establishing a new BRICS currency basket and making greater use of local currencies for trade remain key priorities. It is interesting that Brazil, China, and South Africa show higher potential for currency stability and de-dollarization compared to India and Russia (Bastanifar et al., 2025). Ongoing geopolitical tensions, such as Russia's sanctions and China's BRI, will keep de-dollarization on the radar. If the new Trump administration were to continue engaging in trade wars (through higher tariffs) with especially Eastern countries, it would reinforce their efforts to distance themselves from all dominant aspects of the Western financial system. It could actually play into BRICS+ nations' hands because this growing bloc of nations do have significant transformative potential for reshaping the global financial architecture. Notably, the new US Treasury secretary, Scott Bessent, said the government will use stablecoins to ensure that the US dollar remains the world's global reserve currency (Dale, 2025).

BRICS+ represents a growing challenge to US hegemony by advocating for a multipolar world order, fostering alternatives to US-led economic systems, and building influence in developing regions. Its impact is still moderated, however, by internal divisions and the entrenched dominance of US-centric systems. As BRICS+ strengthens and expands (growing in membership, collective GDP and influence), its role as a counterbalance to US dominance will likely grow, particularly in reshaping global economic and political norms as well as geopolitical dynamics. A last and quite interesting observation is how the past might now be catching up with the West. The historical exploitation of the Global South through Western colonialism and neoliberal policies continue to shape global inequalities, which cause many developing economies to look elsewhere. The West's emphasis on free trade and financial liberalization has often marginalized these economies, while BRICS+ promotes a more equitable global system. The BRICS expansion signals the beginning of a more competitive multipolar world order, with the Global South gaining ever more influence in directing it.

References

Adua, L., Zhang, K. X., & Clark, B. (2021). Seeking a handle on climate change: Examining the comparative effectiveness of energy efficiency improvement and renewable energy production in the United States. *Global Environmental Change, 70*, 1–10.

Archer, A., Thomas, A., & Engelen, B. (2022). The politics of envy. In S. (Ed.) *The moral psychology of envy*. Rowman & Littlefield. https://doi.org/10.5771/9781538160077

Bastanifar, I., Khan, K. H., & Koch, H. (2025). Understanding BRICSIZATION through an economic geopolitical model. *Journal of Open Innovation: Technology, Market, and Complexity, 11*, 1–10.

Bergstresser, D. (2024). *The rising burden of U.S. government debt*. https://econofact.org/the-rising-burden-of-u-s-government-debt#:~:text=There%20has%20been%20a%20notable,required%20increased%20federal%20spending%20and

Bharat, A., Gautam, R. S., & Rastogi, S. (2024). Trends of currencies in Forex reserves: Whither de-dollarization? Conference Proceedings of the 2024 International Conference on Automation and Computation (AUTOCOM) in Dehradun, India. https://doi.org/10.1109/AUTOCOM60220.2024.10486118

Bilal, A., & Rossi-Hansberg, E. (2023). Anticipating climate change across the United States. *NBER Working Paper*, 31323, 1–70.

Bolt, J., & Van Zanden, J. L. (2024). Maddison-style estimates of the evolution of the world economy: A new 2023 update. *Journal of Economic Surveys, Special Issue*, 1–41.

Bulut, U., Ongan, S., Dogru, T., Isik, C., Ahmed, M., Alvarado, R., Amin, A., & Rehman, A. (2023). The nexus between government spending, economic growth, and tourism under climate change: Testing the CEM model for the USA. *Environmental Science and Pollution Research, 30*, 86138–86154.

Caldararo, N. (2023). China, invasions in Europe and de-dollarization: Dismembering empires and collapse of nations. *The IUP Journal of International Relations, 17*(4), 7–40.

Caria, S. (2022). Cooperation regimes and hegemonic struggle: Opportunities and challenges for developing countries. *Politics and Governance, 10*(2), 71–81. https://doi.org/10.17645/pag.v10i2.4919

Carmody, P. (2024). BRICS' enlargement: Power expansion or contraction in a changing world order? *EconPol Forum, 25*(1), 14–17.

Chen, H., Shi, Y., & Zhao, X. (2022). Investment in renewable energy resources, sustainable financial inclusion and energy efficiency: A case of US economy. *Resources Policy, 77*, 1–8.

Congressional Budget Office. (2024). *The distribution of household income in 2021*. https://www.cbo.gov/publication/60341#data

Dale, B. (2025). *Stablecoins will preserve dollar dominance, Bessent says*. https://www.msn.com/en-us/money/markets/stablecoins-will-preserve-dollar-dominance-bessent-says/ar-AA1AtEIS

Deller, S. (2024). *Net savings trends and their impact on the U.S. economy*. https://economicdevelopment.extension.wisc.edu/2024/09/06/net-savings-trends-and-their-impact-on-the-u-s-economy/

Dorn, D., & Levell, P. (2024). Trade and inequality in Europe and the US. *Oxford Open Economics, 3*(1), 42–68. https://doi.org/10.1093/ooec/odad046

Evgenidis, A., & Malliaris, A. (2023). House bubbles, global imbalances and monetary policy in the US. *Journal of International Money and Finance, 138*, 1–16.

Falarti, M. M., & Naqvi, S. F. H. (2024). Eurasia's economic shift: The rise of de-dollarization and the quest for financial autonomy. *Journal of Iran and Central Eurasia Studies, 7*(1), 95–110. https://doi.org/10.22059/jices.2024.98941

Farooq, H., Khan, M., & Uzzaman, M. S. (2020). The 21st century world order: Rise of China and challenges to American global hegemony. *ASIAN Journal of International Peace & Security (AJIPS), 4*(2), 1–12.

Federal Reserve Bank of St. Louis. (2024). *FRED Economic Data.* https://fred.stlouisfed.org/series/GFDEBTN

Firdaus, N., Widayanto, S., & Indrayani, I. (2023). BRICS against hegemony of the United States dollar in the international monetary system. *Journal of Social Political Sciences, 4*(4), 421–442.

Giustra, F. (2023). *De-dollarization: Not a matter of if, but when.* https://responsiblestatecraft.org/2023/05/03/de-dollarization-not-a-matter-of-if-but-when/

Gould, E., & Kendra, J. (2021). Inequality in annual earnings worsens in 2021. *Economic Policy Institute.* https://www.epi.org/publication/inequality-2021-ssa-data/

Herrnstadt, E., & Dinan, T. (2020). CBO's projection of the effect of climate change on U.S. economic output. *Congressional Budget Office Working Paper, 6*, 1–21.

Hilotin, J. (2024). *BRICS Pay vs US dollar vs Euro? What we know so far.* https://gulfnews.com/special-reports/brics-pay-vs-us-d

Hoffman, F., Lee, D. S., & Lemieux, T. (2020). Growing income inequality in the United States and other advanced economies. *Journal of Economic Perspectives, 34*(4), 52–78.

International Monetary Fund (IMF). (2023). *World economic outlook: near-term resilience, persistent challenges (July).* IMF.

International Monetary Fund (IMF). (2024a). *Currency Composition of Official Foreign Exchange Reserves (COFER).* https://data.imf.org/?sk=e6a5f467-c14b-4aa8-9f6d-5a09ec4e62a4

International Monetary Fund (IMF). (2024). *World economic outlook: Policy pivot, rising threats (October).* IMF.

Johnston, M. (2024). *A history of income inequality in the United States.* https://www.investopedia.com/articles/investing/110215/brief-history-income-inequality-united-states.asp#:~:text=Income%20inequality%20is%20a%20problem,mental%20illness%2C%20and%20social%20unrest

Kumar, R., Kumar, P., & Verma, O. M. (2024). Brics new currency: An investigation of trade dynamics to de-dollarizing intra-Brics trade with the reference of Indian bilateral trade contracts. *Journal of Informatics Education and Research, 4*(2), 1376–1383.

Liao, Y., & Walls, M. A. (2024). *Storm watch: Trends in extreme US weather events.* https://www.resources.org/common-resources/storm-watch-trends-in-extreme-us-weather-events/

Liu, Z. Z., & Papa, M. (2022). *Can BRICS de-dollarize the global financial system? Elements in the economics of emerging markets.* Cambridge University Press. https://doi.org/10.1017/9781009029544

Mahoney, J. G. (2023). China's rise as an advanced technological society and the rise of digital orientalism. *Journal of Chinese Political Science, 28*, 1–24.

Matthews, B. (2024). *Report: Average American debt now exceeds $100,000.* https://www.matthewsandmegna.com/posts/report-average-american-debt-now-exceeds-100000?utm_source=chatgpt.com

Mayer, J. (2024). De-dollarization: The global payment infrastructure and wholesale central bank digital currencies. *FMM Working Paper*, 102, Hans-Böckler-Stiftung, Macroeconomic Policy Institute (IMK), Forum for Macroeconomics and Macroeconomic Policies (FMM), Düsseldorf, pp. 1–26.

McKay, L.C. (2023). The state of income inequality. *Federal Reserve Bank of Minneapolis.* https://www.minneapolisfed.org/article/2023/the-state-of-income-inequality#:~:text=Indeed%2C%20income%20inequality%20in%20the,diagnosing%20what%20is%20causing%20inequality

Mohaddes, K., Ryan, N. C., Pesaran, M. H., Raissi, M., & Yang, J. (2023). Climate change and economic activity: Evidence from US states. *Oxford Open Economics*, 1–11. https://doi.org/10.1093/ooec/odac010

NCEI (National Centers for Environmental Information). (2024). *Billion-dollar weather and climate disasters*. https://www.ncei.noaa.gov/access/billions/events#:~:text=In%202024%2C%204%20Tropical%20Cyclone,States%20(CPI%2Dadjusted).

Ng, J. (2024). The BRICS plus challenge and emerging hierarchical multilateralism. *Global Governance: A Review of Multilateralism and International Organizations, 30*(2), 270–287. https://doi.org/10.1163/19426720-03002005

Nikolic, G. V., Lobanova, J. Z., & Zvezdanovic. (2023). Is de-dollarization inevitable? *Kultura polisa, 21*(1), 127–143.

Nwaorgu, O. C., & Nwankwo, C. A. (2015). US hegemony and BRICS' challenge. *Development Studies Roundtable, 5*(2), 1–15.

NWS (National Weather Service). (2024). *Active alerts*. https://www.weather.gov/

OECD (Organization for Economic Co-operation and Development). (2024). *Sustainable Development Goals (SDGs)*. https://www.oecd.org/en/topics/policy-issues/sustainable-development-goals-sdgs.html

Öniş, Z., & Kutlay, M. (2020). The new age of hybridity and clash of norms: China, BRICS, and challenges of global governance in a postliberal international order. *Alternatives, 45*(3), 123–142.

Özekin, M. K., & Sune, E. (2023). Contesting hegemony: The rise of BRICS and the crisis of US-led Western hegemony in the MENA region. *The Korean Journal of International Studies, 21*(3), 409–446. https://doi.org/10.14731/kjis.2023.12.21.3.409

Shah, M. U., Gill, Q. S., Akhtar, M. N., & Mustafa, G. (2024). Understanding the decline of the US dollar: Key factors and implications. *Journal of Social & Organizational Matters, 3*(2), 234–250.

Shao, X., Zhong, Y., Li, Y., & Altuntas, M. (2021). Does environmental and renewable energy R&D help to achieve carbon neutrality target? A case of the US economy. *Journal of Environmental Management, 296*, 1–10.

Siddiqui, K. (2024). The BRICS expansion and the end of Western economic and geopolitical dominance. *The World Financial Review*. https://worldfinancialreview.com/the-brics-expansion-and-the-end-of-western-economic-and-geopolitical-dominance/

Siripurapu, A. (2022). The U.S. inequality debate. *Council on Foreign Relations*. https://www.cfr.org/backgrounder/us-inequality-debate

Srinivasan, H. (2024). *U.S. National Debt by Year*. https://www.investopedia.com/us-national-debt-by-year-7499291

Stiglitz, J. (2013). *The price of inequality: How today's divided society endangers our future*. W.W. Norton & Co.

Thiagarajan, R., Lacaille, R. F., Hurd, A., Metcalfe, M., & Im, H. (2023). *Is US dollar dominance dented?* State Street Corporation.

Türker, H. (2024). Thirty years' crisis: The decline of U.S. hegemony and the rise of China. *Tesam Akademi Dergisi, 11*(2), 557–590. https://doi.org/10.30626/tesamakademi.1337511

U.S. Treasury. (2024). *FiscalData. What is the national debt?* https://fiscaldata.treasury.gov/americas-finance-guide/national-debt/

Vahlsing, C., & Yagan, D. (2022). Quantifying risks to the federal budget from climate change. *White House Briefing*. https://www.whitehouse.gov/omb/briefing-room/2022/04/04/quantifying-risks-to-the-federal-budget-from-climate-change/

Van Nieuwkerk, A. (2024). Brics+ could shape a new world order, but it lacks shared values and a unified identity. *The Conversation*. https://theconversation.com/brics-could-shape-a-new-world-order-but-it-lacks-shared-values-and-a-unified-identity-242308

Warren, K. (2024). *What is the average credit card debt in America?* https://www.investopedia.com/what-is-the-average-credit-card-debt-in-america-8622723?utm_source=chatgpt.com

Wearden, G. (2024). Trump threat of 100% tariffs against Brics nations raises trade war fears. *The Guardian.* https://www.theguardian.com/business/2024/dec/01/trump-threat-100-per-cent-tariffs-brics-nations-dollar

World Bank. (2024). *Gini index—United States.* https://data.worldbank.org/indicator/SI.POV.GINI?locations=US

Zabat, L. H., Sadaoui, N. A., Abid, M., & Sekrafi, H. (2022). Threshold effects of renewable energy consumption by source in U.S. economy. *Electric Power Systems Research, 213*, 1–12.

Competition from the East: The Digital Revolution and How It Affects SDG-Attainment

3

Abstract

The digital revolution is rapidly changing the world, with the US and China competing for technological dominance. This chapter investigates the rivalry between the East and the West in the digital revolution, particularly focusing on technological supremacy in areas like AI, digital payment systems, semiconductors and 5G networks, leading to both innovation and fragmentation. It examines the rise of digital giants in the East, such as China, India, Japan, and South Korea, and their strategies to challenge Western dominance. They are expanding globally and promoting alternative models of digital governance. Despite the intensifying competition, it also highlights the paradoxical interdependence between especially the US and China, which is affecting global geopolitical alignments. The chapter further explores how the digital revolution, driven by the Fourth Industrial Revolution (4IR), impacts the United Nations Sustainable Development Goals (SDGs), presenting both opportunities and challenges. One key finding is that while digital technologies can accelerate progress toward the SDGs through enhanced connectivity, financial inclusion, and smart solutions, they also pose risks such as the digital divide, environmental impact, and ethical concerns. The chapter concludes by emphasizing the need for a techno-ethical framework and inclusive digital transformation to ensure equitable and sustainable development.

Keywords

Global competition • Digital revolution • SDGs • AI • 4IR • Digital transformation

A. J. van Niekerk, *West to East: A New Global Economy in the Making?*,
https://doi.org/10.1007/978-3-031-93267-0_3

1 Introduction

Now that the price of one bitcoin has broken through the US$100,000 threshold and kept rising to over US$120,000 in July 2025, the arrival of cryptocurrencies as an investment and payment alternative is firmly established. New digital payment systems are reshaping the global financial landscape. As much as Donald Trump promised that he is planning a "crypto game-changer" by establishing a "strategic national bitcoin reserve", Russia is considering doing the same, while China and India are also fully invested in the crypto race (Bambrough, 2024). The US Treasury even went as far as calling bitcoin "digital gold" and seeing the potential for bitcoin to eclipse gold's US$16 trillion market capitalization eventually. A bill was introduced in the US congress, entitled the Boosting Innovation, Technology and Competitiveness Through Optimized Investment Nationwide (BITCOIN) Act, recommending that the US buy 1 million bitcoins over five years to reduce the spiralling US$35 trillion US national debt. Russia's President Putin recently signed a new law legalizing bitcoin and cryptocurrency mining in the country and spoke very favourable of bitcoin as an alternative to circumvent crippling financial sanctions (Reuters, 2024). Risking a bitcoin cold war with the US, the Central Bank of Russia is actively preparing to conduct experiments involving cryptocurrency for cross-border settlements.

The escalating competition between the US and China for technological dominance is preparing the ground for an all-out 'tech war' between East and West. This rivalry spans critical sectors such as AI, semiconductor manufacturing, 5G networks, quantum computing, green technology, and digital payment systems (finance). Countries on both sides aim to control critical technologies that drive economic growth, national security, and global influence. The US has, for instance, imposed restrictions on Chinese tech companies, such as Huawei, and limited China's access to advanced semiconductor technologies. China retaliated with similar restrictions on US companies like Facebook and Google. Technological decoupling is taking place, involving efforts to separate supply chains and reduce reliance on one another for critical components and innovation (Mirrlees, 2024). Investments in R&D are growing rapidly to gain a competitive edge. While it accelerates technological innovation, the tech war is creating risks of fragmentation in global standards and supply chains, affecting global trade, investment flows, and the alignment of nations with either bloc (East and West). This rivalry is shaping the future of the global economy and geopolitics, with far-reaching consequences for innovation, security, and international collaboration.

The global expansion of digital imperialism by specifically the US and China divides the world into competing technological spheres of influence. This rivalry shapes digital ecosystems in developing countries, drives geopolitical tensions as nations are forced to align with one bloc or the other, and may exacerbate economic dependencies. Ultimately, both nations use digital tools not only to expand their economic power, but also to project their political and cultural influence on the global stage. This has major implications for economic sustainability and countries' progress towards attaining the SDGs as the digital revolution is offering transformative opportunities but also pose unique challenges. The chapter investigates this aspect, but first examines the landscape and the composition of the digital giants rising in the East, followed by the main 'arenas' where tech competition are taking place.

2 Digital Giants of the East

It is quite possible that the launch of ChatGPT by OpenAI in November 2022 officially kicked off, not just an AI arms race, but a 'tech battle' between East and West the likes of which the world has never seen before. Since its launch, ChatGPT has become a worldwide sensation. Millions have used it for writing poetry, developing apps, and even engaging in informal therapy sessions. It has been adopted—though with varying degrees of success—by news organizations, marketing agencies, and business executives. Additionally, it has sparked a rush among investors eager to capitalize on the next wave of the AI revolution. All this happened predominantly in the West. However, over the past decade, a gradual and much intensified effort by countries and companies in the East has edged ever closer to the West's tech dominance—if not overshadowing the West in certain areas and industries.

China's government has set ambitious goals to become the world leader in AI by 2030 through state-led initiatives and investments (e.g., the New Generation AI Development Plan and companies like ByteDance). Apart from other major tech initiatives, China is actively engaging in international technology governance to influence global standards and policies. By guiding global discussions and organizations, China seeks to shape the technological landscape in ways that align with its strategic interests. India has also articulated a multifaceted strategy to establish itself as a global technology leader, prioritizing digital infrastructure, advanced technology and innovation (e.g., the National Quantum Mission, the India Semiconductor Mission, and The Indian Space Research Organisation), new AI and data centres, and deep tech sector development. Taiwan is of course the global leader in semiconductor manufacturing and chip production, playing a critical role in the global technology supply chain. Japan is actively pursuing several strategic initiatives to solidify its position as a global technology leader. These include the Society 5.0 Initiative, strong leadership in the 4IR, revitalizing the semiconductor industry (e.g. Rapidus, a government-backed semiconductor manufacturer), AI advancements to integrate AI into the workplace to enhance productivity, and major innovations in clean energy technology. South Korea identified a number of areas to establish itself as a global tech leader, such as: developing its high-tech industry (it invested over US$500 billion since 2023 in semiconductors, flat screen displays, future vehicles, biopharmaceuticals, and robotics), a 2030 National Strategy for Artificial Intelligence, digital talent development programmes, strengthening intellectual property protection (e.g., developing a big data system at the Patent Agency), and human-centred technology innovation (e.g., startups in quantum computing and blockchain technologies in regions like Pangyo) (Dong-Yeon, 2025).

This is indicative of a new type of geopolitical rivalry in the digital age, where the competition for technological dominance, especially in AI, is a major factor shaping the global landscape. While the US tech giants initially had a first-mover advantage, Chinese companies like Alibaba and Tencent have emerged as serious contenders, challenging the existing order and prompting a strong techno-nationalist response from the US (Lundvall & Rikap, 2022). China has made significant strides in developing its domestic technological capabilities, particularly in AI, and becoming the only country where US tech giants have a low market share. This rise is attributed to:

- *Government support*: China has declared AI a strategic technology and is actively investing in its development.
- *A strong domestic knowledge base*: China's focus on building a strong domestic knowledge base has allowed it to negotiate the openness of its innovation system.
- *A different division of labour between universities and tech giants*: The specific collaboration between universities and tech giants in China has contributed to the development and application of AI (Hemmings, 2020).
- *Access to the gigantic Chinese market and data*: Chinese tech giants have privileged access to the large Chinese market and its data, fuelling their growth.

The US responded to this by adopting a more protectionist stance. It implemented a multifaceted strategy, aiming to preserve its leadership in critical technology sectors and ensuring national security (Mearsheimer, 2021). The US's protectionist strategy is marked by:

- *Techno-nationalism*: The US government has increasingly focused on protecting its technological advantage and hindering China's progress. The US has undertaken policy reforms to address the challenges posed by China's technological advancements. This includes rethinking technology transfer policies and implementing strategies to reduce dependence on Chinese suppliers (Lewis, 2023). Efforts are also being made to harmonize state capacities in the high-tech realm to match China's robust technology policies.
- *Sanctions and restrictions*: The US has imposed sanctions and restrictions on Chinese tech companies, citing national security concerns. The US government has imposed strict export controls to limit China's access to advanced technologies, predominantly in the semiconductor and AI sectors. For instance, recent measures restrict the export of advanced computer chips essential for AI development to China (Perrigo, 2025). These actions aim to prevent China from enhancing its AI capabilities, which could have significant economic and military implications.

- *Investment in domestic industries (critical technologies)*: The US is investing heavily in its domestic semiconductor industry and other strategic sectors to maintain its technological leadership. In the 2022 fiscal year, the federal government allocated nearly US$200 billion to sectors such as biotechnology, quantum computing, and advanced manufacturing. Despite these investments, reports indicate that China is advancing rapidly, often outpacing the US in patent filings and technological developments (DiNardo, 2023).
- *Emphasis on cybersecurity and digital diplomacy*: The US has made significant efforts to safeguard its technological infrastructure, taking cybersecurity and digital diplomacy to new heights. Officials have advocated for proactive engagement in global tech and digital security arenas to counter the influence of nations like China and Russia (Geller, 2025). This includes promoting Western technology standards and supporting allies with tech investments to ensure a secure and resilient digital ecosystem.
- *International collaborations and alliances*: To counter China's technological rise, the US has strengthened alliances and collaboration with key partners, such as Japan and the Netherlands. This has led to coordinated export controls, especially concerning semiconductor manufacturing equipment. These alliances aim to restrict China's ability to procure essential technologies and components necessary for its tech industry.

This rivalry is reshaping the global innovation landscape and has several global implications. The competition is leading to a more fragmented internet, with countries aligning themselves with either the US or China. As the digital divide increases, smaller economies face difficulties in catching up and are forced to reconsider their openness in terms of data, foreign acquisitions, and scientific collaboration (Mirrlees, 2024). The intensifying competition raises concerns about a potential digital Cold War, dividing the world into separate spheres of knowledge production and use, resulting in increasing fragmentation of global supply chains. In this way, parallel technological ecosystems are developing and a technological decoupling takes place (both nations reduce reliance on each other's technologies). This could lead to regional blocs of technological dominance, mirroring the ideological divisions of the original Cold War (Mearsheimer, 2021). This could lead to restricted cross-border knowledge sharing, slowing down global innovation. When concerns grow about cyber threats, espionage, data breaches, and misinformation campaigns, there could be further deepening mistrust between the two sides. A digital Cold War could force multinational companies to choose sides, leading to market fragmentation and higher costs for consumers. Companies that rely on global supply chains and markets might have to navigate conflicting regulatory environments, increasing complexity and operational challenges.

Tech companies in the East (mainly China, Japan, South Korea, and India) that have become global powerhouses in sectors such as e-commerce, social media, AI, and consumer electronics are quite a number. The Alibaba Group, known as the "Amazon of China," dominates online retail in China and operates cloud services under Alibaba Cloud. It specializes in e-commerce, cloud computing, and FinTech. Another Chinese tech giant is Tencent, which specializes in social media, gaming, and FinTech. It is the creator of WeChat, China's most popular social app, and is a major player in the global gaming industry, owning stakes in companies like Epic Games and Riot Games. Huawei, specializing in telecommunications, AI, consumer electronics, is a global leader in 5G technology and smartphones, despite facing geopolitical challenges and trade restrictions from the US (Mirrlees, 2024). Baidu, often called the "Google of China", leads in AI-driven search technology and autonomous driving research, specializing in internet searching, AI, and autonomous vehicles.

Japan's tech giants have a long-standing reputation for innovation in robotics, consumer electronics, and gaming. Sony is a global leader in gaming (PlayStation), imaging sensors, entertainment, and consumer electronics. SoftBank, known for their tech investments and role in AI and telecommunications, is a major investor in specifically global tech startups and AI initiatives (e.g. The Stargate Project). Panasonic specializes in smart city solutions, the IoT, batteries (energy solutions), and consumer electronics. Nintendo is a pioneer in digital entertainment and the gaming industry with globally recognized franchises like Mario and Zelda.

South Korea is a hub for electronics and digital innovation, with strong government investment in technology and a tech-savvy population. Two well-known tech giants are Samsung and LG Electronics. Samsung, specializing in consumer electronics, semiconductors and AI, is a global leader in smartphones, memory chips and display technology. LG Electronics specializes in smart home solutions (consumer electronics), AI-driven appliances (e.g., dishwashers, refrigerators, washing machines, home theatre systems, and all kinds of smart appliances), and OLED display technology (it is the largest shareholder of LG Display, the world's largest display company by revenue in 2024, and the world's second-largest TV manufacturer). Other tech giants in South Korea include Naver, known for being the country's dominant search engine, and a key player in AI development and cloud computing. Kakao is another fast-growing company, specializing in messaging, payments, AI services, and known for its popular messaging app KakaoTalk. It is also expanding strongly into FinTech.

India is home to several global tech giants that have significantly influenced the international technology landscape. Prominent among these are first, Tata Consultancy Services, India's largest IT services firm, which offers services such as consulting, software development, and business solutions to clients worldwide. Secondly, Infosys provides IT consulting and services globally, specializing in areas such as AI, cloud computing, and digital transformation. Thirdly, HCLTech delivers IT services and products across various sectors, including healthcare, finance, and aerospace, with a strong international presence. Fourthly, Wipro offers IT services, consulting, and business process outsourcing, serving clients across multiple industries worldwide. Fifthly, Tech Mahindra specializes in digital transformation, consulting, and business re-engineering services, operating in numerous countries and catering to various industries. Lastly, Cognizant; although headquartered in the US, it has significant operations in India and is considered part of the

Indian IT giants. It provides IT services, including digital, technology, consulting, and operations services. Together, these tech companies are collectively referred to as "TWITCH" and known in the media as the "Big Six" (Glover, 2025a). These firms are among the dominant players in the global delivery model, providing IT consulting, outsourcing and offshore software development to Fortune Global 500 corporations and government entities. Featuring among the 20 largest employers in India, they have been instrumental in positioning India as a global hub for IT services.

All these tech companies have experienced explosive growth, especially in the last decade, and have become major players in the digital economy and serious AI competitors. They are "national champions" and global pioneers for a number of reasons, including (Lundvall & Rikap, 2022):

- *Large domestic markets*: countries like China and India have massive populations, creating strong demand for digital services and products, with a rapidly growing pool of data. These companies dominate the Asian digital services market, and have privileged access to the large Chinese, Indian, Japanese, Singaporean and Indonesian markets and their data.
- *Government support*: policies such as "Made in China 2025" and South Korea's investment in 5G are significant stimuli to digital growth. The Chinese government, for instance, has designated companies like Alibaba and Tencent as "national AI team" members, giving them key roles in building China's AI-powered future.
- *Technological innovation*: heavy investment in AI, robotics, and semiconductors has made these companies global leaders. In terms of platform innovation, China has emerged as a world leader in building super apps and super platforms. All the leading Asian countries are heavily investing in various kinds of AI research and development. A Report by Govini (2023), a data analytics firm, found that in every critical technology subsegment, China has surpassed the US in granted patents as of 2022. It found that the difference in the pace of technological advancement between the US and China is largely due to accessibility to patents. According to Govini's chairperson and former deputy defence secretary, Bob Work, "Patents are a leading indicator of technological dominance in the future. [Patents] are the seed for making new discoveries that put you on the top of the competitive food chain. … That's what scares me the most [because] China's doing far better than us in terms of the overall number of patents. … If we don't invest further, from semiconductors, to overhead capabilities, to [artificial intelligence], quantum analytics, to biology, advanced energy, those are domains where if China dominates, that will pose as much of a national security threat, both in terms of their ability to spread their influence, or frankly, to take us offline, as anywhere else" (DiNardo, 2023).
- *Global expansion strategies*: Many Eastern digital giants are expanding into emerging markets in Asia, Africa, and Europe. These tech companies are expanding globally, particularly in developing countries, often leveraging their

experience in building digital infrastructure at home. Eastern digital giants, primarily from China, South Korea, India, and Japan, are leveraging their strengths in innovation, competitive pricing, and strategic partnerships to expand globally. They are making full use of their technological expertise, financial strength, strategic investments, and government support to establish a strong presence across international markets. Their approach focuses on a mix of emerging market dominance, strategic acquisitions, and localization efforts to compete with Western tech firms.

- *Vertical integration*: Companies like Alibaba and Samsung have diverse portfolios, spanning multiple industries from FinTech to consumer electronics. Eastern tech giants strongly prioritize gaining full control of entire value chains, from production to distribution and services—shielding them from geopolitical disruptions. This approach allows them to achieve cost efficiencies (lower reliance on third parties), faster innovation cycles, and tighter control over data ecosystems (unmatched access to user behaviour and preferences), giving them a strong competitive edge in global markets. It enables faster market expansion as integrated ecosystems provide a comprehensive solution that attracts and retains users.

It is interesting that, in 2023, the US tech giants known as The Magnificent Seven collectively returned 111% on an equal-weighted basis. Together, they accounted for nearly 60% of the S&P 500's return. In contrast, the Chinese tech sector lagged in 2023, falling 8%. In 2024, the Magnificent Seven—Apple, Microsoft, Amazon, Alphabet (Google), Nvidia, Meta, and Tesla—gained 63%, accounting for about 75% of S&P 500 earnings growth (Laidley, 2025). Collectively, the Magnificent Seven generate more profit in a single year than the total market capitalization of most countries. Even within the US, their earnings far surpass those of other companies. Without these seven giants, the overall earnings per share of the S&P 500 would have declined in 2023 instead of showing growth. The ongoing AI surge has been the primary driver behind the continued rise in the stock prices of these US tech giants. Like their US counterparts, Chinese tech companies benefit from similar long-term growth trends (FOORD, 2024). For instance, Tencent parallels Facebook in digital advertising expansion, JD and Alibaba mirror Amazon's growth in e-commerce, and Baidu aligns with Google and Nvidia in the advancement of AI. However, a significant difference lies in their market focus—while US tech companies have a truly global presence, Chinese tech firms remain primarily concentrated on their domestic market. Table 1 provides a summary of the growth comparison of the leading US and Chinese tech giants. While the size and growth rates of the US firms are on average higher than their Eastern counterparts, the latter have shown strong growth as they keep expanding globally.

It is important to note that this competition is not just between companies but also between states, intertwined with geopolitical strategies and national security interests. The outcome of this rivalry will have a profound impact on the global economic and political order in the years to come. Despite intense rivalry, it is worth emphasizing that the US and Chinese economies are deeply interconnected,

Table 1 Comparing leading tech firms in the US with those in the East (in US$)

Rank (2025)	Magnificent seven (US)	Market cap 2015	Market cap Jan 2025	% growth	Tech firms in the East	Market cap 2015	Market cap Jan 2025	% growth
1	Apple Inc	$598bn	$3.1 trillion	418	TSMC (T)	$150bn	$1.2 trillion	700
2	Microsoft	$440bn	$3.0 trillion	581	Tencent (C)	$190bn	$0.4 trillion	110
3	Nvidia	$15bn	$2.9 trillion	19,233	Samsung (SK)	$177bn	$0.3 trillion	70
4	Alphabet	$534bn	$1.9 trillion	87	Alibaba (C)	$212bn	$0.2 trillion	– 6
5	Amazon	$323bn	$1.7 trillion	426	JD.com (C)	$46bn	$0.1 trillion	117
6	Meta	$273bn	$1.0 trillion	266	PDD Hol-dings (C)	$24bn	$143 billion	495
7	Tesla	$30bn	$0.9 trillion	2900	Meituan (C)	$52bn	$118 billion	127

Note In the tech firms in the East column, *T* Taiwan, *C* China and *SK* South Korea
Sources Kochkokin (2017), Laidley (2025), Statista (2025), Wendling (2024), Genspark (2024)

and complete disentanglement is unlikely. Furthermore, both the US and China are trying to influence the global governance of technology through international organizations and partnerships, each aiming to promote their own digital tech models and standards (Mirrlees, 2024). This competition extends beyond the economic and technological realms, influencing public discourse and international relations. Yet, as stated, within this rivalry, there is continued collaboration and competition between US and Chinese companies, driven by profit motives. The future of this complex relationship remains uncertain and will depend on the strategic choices made by both states and corporations. The one factor, though, which is propelling Eastern tech giants to excel in outperforming their Western counterparts, is vertical integration. It is through:

- *Cost control*: lower reliance on third parties results in reduced costs and higher margins.
- *Faster innovation*: in-house production enables quicker adaptation to market changes.
- *Data dominance*: end-to-end ecosystems provide unparalleled access to user behaviour and preferences.
- *Resilience*: greater control over supply chains shields Eastern companies from geopolitical disruptions.
- *Market expansion*: integrated ecosystems provide an all-in-one solution that appeals to users, and retains them.

One area that has been problematic to Eastern tech companies in expanding globally has been regulatory challenges in the West. In the past decade, they have actually done well in overcoming many of them (Tirole, 2023). They have followed a number of strategies. Firstly, they learned to adapt to different regulatory frameworks (e.g., General Data Protection Regulation (GDPR) in Europe and California Consumer Privacy Act (CCPA) in the US), allowing Eastern tech companies to enter and operate in highly regulated markets. Companies like Alibaba and Tencent adjust their data privacy policies and business operations to comply with international laws, facilitating smoother market entry. Many Eastern firms have also established regional headquarters in key markets to comply with local regulations. Secondly, they have intentionally built trust and credibility in the West. Compliance with local regulations helps companies improve their public perception and build trust with consumers and governments. For example, Huawei, facing scrutiny over security concerns, has implemented greater transparency initiatives to reassure stakeholders in Europe and Africa. This enabled them (and other Eastern firms) to implement rebranding strategies. Huawei started to focus on non-sensitive product categories in Western markets. They are also investing in corporate social responsibility (CSR) initiatives to build goodwill and trust.

Thirdly, through strategic partnerships and alliances, these firms have formed joint ventures and partnerships with local companies, easing entry into foreign markets. TikTok (Bytedance) has partnered with local regulatory bodies to address concerns about data storage and content moderation in, for instance, the US and

India. In terms of corporate culture, the Chinese companies are known for being more adaptive in their approach to different markets and regulatory frameworks (Erie & Streinz, 2021). Fourthly, adapting to foreign regulations often encourages diversification of products and services to meet compliance standards. An example is Alibaba Cloud, which has expanded its infrastructure to comply with European data residency requirements, boosting cloud services globally. Sixth, as they expand, Eastern digital giants are constantly engaging with policymakers and industry associations in the West to shape favourable regulations in new markets. For instance, participation in regulatory discussions have helped companies like Baidu align their AI developments with global ethical and legal standards.

Eastern tech companies, particularly those from China, are playing a significant role in reshaping the digital order, challenging the dominance of Western, especially US-based, tech companies and their associated norms (Mirrlees, 2024). This is happening through several key mechanisms:

- *Building digital infrastructure*: Chinese companies are actively constructing digital infrastructure globally, especially in developing countries, through projects like the Digital Silk Road (DSR). This includes laying fibre optic cables, building data centres, providing smart city solutions, and establishing mobile networks (Cheng & Zeng, 2024). By providing the physical and digital components of the internet across Asia, Africa, and Latin America, they are shaping the conditions under which countries transition towards digital societies and present an alternative to Western-led infrastructure like AWS and Google Cloud. Huawei and ZTE are, for example, leading 5G deployments worldwide, positioning themselves as critical enablers of the global digital economy.
- *Developing alternative digital ecosystems*: Tech companies from especially China, South Korea, and India are creating self-sustaining digital ecosystems that rival Western platforms. Chinese companies are exporting not just technology but also a 'digital stack' that is comprised of a network infrastructure, devices, applications, content, and governance models. This enables them to exert greater influence over the digital ecosystems of recipient countries (Economy, 2024). These new ecosystems offer consumers and businesses an all-in-one experience, reducing dependence on Western platforms such as Google and Amazon. Examples are Alibaba and Tencent, who have built comprehensive ecosystems, integrating e-commerce, payments (Alipay, WeChat Pay), cloud computing, and AI. Eastern companies are reshaping digital experience through super apps, offering a one-stop solution for communication, commerce, and entertainment. WeChat and Alipay combine messaging, social media, shopping, and banking, setting a precedent that Western firms are now trying to emulate. These platforms introduce an integrated lifestyle, reducing the need for multiple apps and encouraging user stickiness. What is more, these platforms are not just commercial entities, but also vehicles for promoting Chinese standards, payment systems (like the renminbi), and potentially, governance norms.
- *Challenging Western dominance*: The expansion of Chinese tech companies and the adoption of their technologies are challenging the long-standing dominance

of US-based digital platforms and their associated norms (Mearsheimer, 2021). This competition is leading to a more multipolar digital world, where the US is no longer the sole centre of digital power. Eastern tech giants are striving to dominate key emerging technologies by leading in cutting-edge technologies like AI, 5G, blockchain, and smart manufacturing, offering solutions tailored to local markets. Baidu, Tencent, and Alibaba are, for example, at the forefront of AI development, including facial recognition and smart city technologies, while TSMC dominates global semiconductor production, which is critical to digital transformation worldwide (Mirrlees, 2024). Eastern countries are also driving the development of digital currencies and cashless economies, reducing dependence on Western financial systems such as Visa and Mastercard. China's Digital Yuan (e-CNY), promoted by Alibaba and Tencent, and India's Unified Payments Interface (UPI), are efforts to challenge the global dominance of the US dollar in digital transactions by offering frictionless alternatives.

- *Championing alternative tech governance models*: Eastern companies are promoting state-centric and controlled models of internet governance, in contrast to the Western open and decentralized model. China is advocating for a concept of 'data sovereignty', which emphasizes governmental control over data flows (Erie & Streinz, 2021). This conflicts with the US approach that often emphasizes the free flow of information. This concept is appealing to some developing countries who seek greater control over their digital spaces. Countries adopting Chinese technology often follow China's model of digital governance, which includes tighter regulations on data flow and cybersecurity. An example is the 'Great Firewall' approach, which influences digital policies in emerging economies seeking tighter content control and cybersecurity measures (Bremmer, 2021). Chinese companies, particularly Huawei, are increasingly influential in setting global technology standards, especially in 5G. This allows them to embed their technologies and norms into the global digital architecture.

Fears are growing that China's model of digital development, which combines economic growth with strict state control, may be exported to other countries (Gertz & Evers, 2020). The concern is that such 'digital authoritarianism' will lead to a decline in digital freedoms. It is clear that the digital order is undergoing a significant transformation, with Eastern tech companies, particularly those from China, playing a major role. They are reshaping the global digital landscape through innovation, strategic expansion, and the promotion of alternative digital ecosystems. This is not simply a technological shift, but a geopolitical one, with implications for the balance of power and the future of the internet. It is a complicated picture, though, as the relationship between China and its tech companies is not entirely one-sided, given the competitive dynamics that exist within China as well. The outward, global effect is highly significant too as some countries may emulate aspects of China's model, while others may seek to balance their relationships with both the US and China. By overcoming regulatory challenges, focusing on local needs, and investing in innovation, Eastern tech companies are reconstructing the digital order. They are no longer just following Western models but are creating unique, self-sustaining ecosystems that rival global incumbents, positioning themselves as the architects of the next digital era.

At the heart of it all, it is China's strategic objectives for the Digital Silk Road that stem from centralized decision-making, driven by an ideological foundation that fuels the intense on-the-ground efforts. As highlighted by Hemmings (2020: 20), "China's ambitions are of the grandest scale and involve an attempt at multi-regional reordering and, by proxy, global reordering." The integration of ICT with infrastructure has proven to be a winning strategy for China. This approach has enabled Beijing to advance its own standards, companies, and digital currency, providing access to new captive markets for Chinese tech firms, abundant data for analysis, and tools to exert influence over foreign political and business elites. The speed and intensity with which the digital technologies of tomorrow are being created in the East, are shaping the pillars of hegemonic order today. This is setting off a chain reaction. As an increasing number of state power pillars—military, economic, and political—rely on the digital realm for their functionality, governments will leverage these digital assets for geopolitical purposes.

Perhaps the key question for the West is: Will the US and its allies be able to counter the growing dominance of China's technical standards and governance model in shaping the markets and political systems of the future? This question underscores the magnitude of the ideological challenge facing Western nations. Addressing China's authoritarian approach to data, media, and state-citizen relations is of critical importance. In 1946, Winston Churchill's Iron Curtain speech in the US highlighted the Soviet Union's coordinated efforts across Eastern Europe to expand its influence and control. Similarly, the United States and its partners must evaluate China's seemingly disconnected actions—especially through its digital giants—to uncover its broader intentions and anticipate the consequences for those who live under China's 'digital curtain' (Hemmings, 2020). Given his strong, interventionist leadership, could Donald Trump be the Winston Churchill of today? China is certainly here to stay and the tech giants of the East seem to be only in their infant stage. With data becoming the new gold in today's surveillance economy and the key asset for global reordering, the 'tech battle' might only be at its inception.

3 Global Tech Competition and the 'Tech War'

A defining characteristic of this era is the rise of the People's Republic of China and the strategic competition that comes with it. The contest between US and Chinese digital imperialism is leading to the creation of new 'digital spheres of influence,' with countries aligning with either the US or China (Mirrlees, 2024). Smaller countries face difficulties in navigating this new landscape and may become reliant on either US or Chinese technologies, potentially undermining their own autonomy. For example, Eastern companies are at the forefront of a power shift away from legacy media towards digital media companies. The rise of super apps is changing how companies are structured and how they approach business. Furthermore, as tech giants from both the East and West compete for market dominance and user attention, their strategies often target younger audiences, leading to increased screen time and dependency. Clearly, the ripple effects of global tech competition is getting wider and wider, affecting every fabric of society.

The tech competition between the East (mainly China and to some extent South Korea, India and Japan) and the West (led by the US and Europe) is a multifaceted rivalry spanning economic, political, and technological domains. It revolves around dominance in emerging technologies such as AI, 5G networks, semiconductors, quantum computing, e-commerce, cybersecurity, green technology, digital governance, and more. This competition is shaping the future of global technological leadership, economic influence, and without a doubt, geopolitical power. The rivalry between the United States and China is polygonal and complex, stemming from their distinct histories, state systems, economic structures, and national cultural identities (Economy, 2024). Despite decades of collaboration, particularly in the economic realm, the two countries are increasingly engaged in an asymmetrical rivalry across economic, military, and media-technological domains. Now that Donald Trump has become the US President again, with tech leaders like Elon Musk (SpaceX), Jeff Bezos (Amazon), Mark Zuckerberg (Meta), Sandar Pichai (Google), and Tim Cook (Apple) by his side at the inauguration, the tech competition with the East is expected to intensify exponentially, especially given Trump's declaration of a "golden era" for the US (Dolan, 2025). Hours after Trump's inauguration, Chinese President Xi Jinping and Russian President Vladimir Putin had a video call, pledging to bring bilateral ties to "greater heights" (Ng, 2025). Xi called on Putin to "continue deepening strategic coordination, firming up mutual support, and safeguarding legitimate interests". With trade between the two countries having reached a record US$240 billion in 2023—up more than 64% since 2021—their collaboration with regards to competing globally (together with BRICS+ partnerships) is expected to increase. Innovation and technology will no doubt be central to their strategy.

The global expansion of digital imperialism by the US and China occurs through strategies that aim to dominate digital technologies, infrastructure, and governance, leveraging these as tools for economic and geopolitical influence. For decades, there were uncertainties about China's path, its local political and economic reforms, and its ultimate aspirations as a global power. Would it aim to challenge or reshape the liberal international order, or would it settle into a role as a status quo power? Under Xi Jinping, this question has become more pronounced as China has shifted its policies, rhetoric, and objectives from Deng Xiaoping's "hide and bide" strategy to Xi's approach of "moving to center stage" (Hemmings, 2020: 7). One of the most noticeable elements of this shift is the Belt and Road Initiative (BRI) with its strategic significance in specifically connecting the Asian, African, European and South American continents. It is a very intentional attempt to create "a global information highway with China at its core," effectively developing its own sticky power through technology packages that include e-commerce, smart cities, smart ports, digital currency, communications networks, and satellite networks (Shen, 2018: 2692). At its core, this is a gradual, long-term effort to establish an order that aligns more with China's preferences, initially at the regional level and, by extension, on a global scale.

A prime example of the intensifying tech competition between China and the US is China's launch of DeepSeek (its first free chatbot app version) on 20 January 2025, the very day on which Donald Trump was inaugurated as the 47th US President (Ng et al., 2025). By January 27, it had surpassed ChatGPT to become the most-downloaded free app on the iOS App Store in the US, triggering an 18% drop in Nvidia's share price (US$600 billion fall in market value), the largest single-day loss in US history. DeepSeek closely resembles ChatGPT in appearance, functionality, and user experience. As an open-source large language model (operating as a reasoning model, generating responses incrementally to mimic human problem-solving processes), it is powered by the DeepSeek-R1 model for Apple iOS and Android. Founded in 2023, the company is headquartered in Hangzhou, Zhejiang, and owned by a market leader in Chinese hedge funds, Liang Wenfeng, who also serves as its CEO. DeepSeek's rapid rise against larger, more established competitors has been described as a disruption to the AI landscape, marking "the first shot at what is emerging as a global AI space race", and ushering in "a new era of A.I. brinkmanship" (Roose, 2025: 1). Following its launch, DeepSeek quickly topped the Apple Store's download charts, surprising investors and causing a number of tech stocks to tumble.

DeepSeek immediately caught the attention of AI experts, the broader tech industry, and the world. President Donald Trump called it a 'wake-up call' for American companies, urging them to prioritize competition (Ng et al., 2025). Serious concerns have, however, been raised about data privacy and security. One part of DeepSeek's privacy policy states that data can be used to help match the user and his/her actions outside of the service. This rings an alarm bell for anyone concerned about his/her privacy. As its own privacy policy states, DeepSeek collects large amounts of personal information collected from users, which is then stored in what it says is "secure servers in China" (Gerken, 2025). It harvests data like email addresses, phone numbers and dates of birth, which are entered when creating an account, plus any user input including text and audio, as well as chat histories and technical information on the model of phones used, the operating system to a user's IP address and 'keystroke patterns'.

While other similar platforms may ask for the same info, the situation is different with China, being a centralised control regime, using a social credit score system to apply social control. Seemingly innocent, it states that this data is used to enhance DeepSeek's safety, security, and stability. The information may then be shared with entities such as service providers, advertising partners, and its corporate group, where it will be retained for as long as needed. US officials have raised concerns about national security, according to new White House press secretary Karoline Leavitt. The US navy has reportedly issued a complete ban on its members using DeepSeek's apps out of security fears and ethical concerns (Galvin, 2025).

What sets DeepSeek apart is its ability to rival industry-leading models like that of OpenAI and Gemini (Google) at a fraction of the cost by merely relying on fewer advanced chips. It is already as powerful as OpenAI's o1 model—released at the end of 2024—in tasks such as mathematics and coding. It consumes less memory than its competitors, significantly lowering operational costs. According to its developers, DeepSeek was trained for just US$6 million, a fraction of the "over US$100 million" OpenAI CEO Sam Altman hinted at for GPT-4 (Nazzaro, 2025). It appears that DeepSeek has been trained using reliable sources while incorporating an additional safeguarding layer that censors or withholds certain information. Reports suggest that DeepSeek's founder stockpiled Nvidia A100 chips, which have been banned from being exported to China since September 2022 (Ng et al., 2025). It is believed that this stockpile (around 50,000) enabled the development of such a powerful AI model by combining these high-performance chips with more affordable, less advanced ones. DeepSeek challenges Washington's strategy to curb China's tech ambitions, especially its ban on exporting advanced chips to Beijing. China, though, has intensified its AI push, with President Xi Jinping making it a national priority. Start-ups like DeepSeek play a highly strategic role as China shifts from traditional manufacturing industries, such as textiles and furniture, to cutting-edge sectors like AI, electric vehicles, and semiconductor technology.

Equally, the US launched the Stargate Project the day after, on 21 January 2025, in a bid to solidify the US's lead in artificial intelligence. Also known as Stargate LLC, it is an AI joint venture created by OpenAI, SoftBank, Oracle, and an Emirati investment firm, MGX. The initiative aims to invest up to US$500 billion in AI infrastructure across the United States by 2029. Originally conceived in 2022, it has been compared to the Manhattan Project and the Apollo program. President Donald Trump officially announced the venture the day after his second inauguration, accompanied by Sam Altman from OpenAI, Larry Ellison from Oracle and Masayoshi Son, CEO of SoftBank, who is serving as its chairman. Trump called Stargate "the largest AI infrastructure project in history", and suggested that he would use emergency declarations and executive orders to fast-track the project's development, especially regarding energy infrastructure (Gedeon, 2024: 1).

The "Stargate" term in AI context started in 2024 when OpenAI and Microsoft began working on a US$100 billion AI supercomputer. Ten data centres are being constructed in Abilene, Texas, with plans to expand to 20 locations of about half a million square feet each in additional states. The project is expected to generate over 100,000 jobs in the US, hiring people for construction, operations, and other areas. SoftBank, a Japanese company, will oversee the venture's financial responsibility, while OpenAI will handle its operational responsibility (Sherry, 2025). Key initial tech partners include Arm, Microsoft, Nvidia, Oracle, and OpenAI. One focus area, for instance, is AI's ability to transform health care. Tools provided by OpenAI and Softbank are already helping to develop a cancer vaccine through this project. AI can look at a blood test and do early cancer detection that is seriously threatening a person. Then it can gene sequence the gene tumour so that the person can be vaccinated against that cancer. This will make it possible to make that mRNA vaccine robotically using AI within 48 h (Picchi, 2025).

Yet, Stargate appears to prioritize building infrastructure to support individual companies rather than directly financing specific breakthrough research projects. The focus is more on commercial applications than government efforts. According to Trump, "Stargate will be building the physical and virtual infrastructure to power the next generation of advancements in AI" (Glover, 2025b). Importantly, the Stargate Project signals the Trump administration's drive to promote investment and expansion in the AI sector to strengthen the US's AI capabilities amid its competition with China for technological dominance. This is how Wall Street interpreted it, especially after Trump, contrary to his predecessor, repealed within hours after returning to the White House, Biden's policy constraints on AI as a fast-developing technology. Given SoftBank's involvement, it also underscores that Trump is not entirely opposed to foreign firms participating in America's push for AI dominance—as long as they are headquartered in nations that are strong US allies.

In essence, the tech war between the East and the West refers to the growing competition between global tech powers, primarily China and the US, over leadership in various kinds of emerging technology. It represents a struggle for economic influence, technological supremacy, and global governance in the digital age. Key elements of the tech competition can be summarized as:

- *Technological dominance*: Specifically through its BRI and the Digital Silk Road, China is rapidly advancing in AI, 5G (via companies like Huawei), semiconductors, green tech, quantum computing, and digital infrastructure, seeking to shape the global digital order and growing its influence over global standards. The US remains dominant in software, internet platforms, and semiconductor innovation, with companies like Apple, Google, and Microsoft leading in AI, cloud computing, and data management. The race for technological superiority in all areas of advanced digital technologies is at the heart of the tech Cold War. Both countries recognize that leadership in these technologies translates to economic strength, military power, and global influence.
- *Trade and security issues*: The US has imposed sanctions on Chinese tech giants (e.g., Huawei and ZTE), citing national security risks related to espionage and intellectual property theft. China has retaliated by pushing for greater self-reliance, developing its own tech ecosystem, and limiting Western companies' access to its market. The US is placing restrictions on Chinese investment in US tech companies and is also restricting the export of key technologies to China. China is also retaliating by limiting sales of critical materials. The tech war is intertwined with the trade war between the US and China, with both countries imposing tariffs and restrictions on each other's tech industries. Both countries are concerned about the security of their supply chains, and are seeking to reduce their reliance on each other for critical technologies and components; hence, they are adopting more and more techno-nationalist policies. As such, both the US and China are supporting their 'national champion' tech companies, providing them with subsidies, regulatory advantages, and preferential access to their domestic markets

- *Global influence and digital sovereignty*: China advocates for a state-controlled digital model, with strict regulations and data sovereignty laws. The West, led by the US, pushes for open internet governance and less state intervention in technology, aiming for a more free-market approach. While China promotes a vision of 'cyber sovereignty,' where governments have the right to control the internet within their borders (i.e., developing technologies that could be used to monitor and censor the internet), the US presents itself as the defender of an open and free internet, while it seeks to curtail the spread of what it labels 'authoritarian technology' (Bagwandeen, 2021). The ability to set global technology standards is crucial, particularly for data privacy, cybersecurity, and digital trade. Both the US and China are vying to have their standards adopted internationally, which would give their companies a competitive edge.
- *Alliances and partnerships*: The US is working to rally its allies to counter China's technological rise, while China seeks partnerships in the global South. The EU is also trying to carve out a space for itself, but lacks the scale of the two superpowers. Taiwan is a major point of tension, especially given its importance in semiconductor manufacturing. The US is trying to prevent China from gaining control over Taiwan, while China sees Taiwan as part of its territory. Both China and the US are attempting to build 'digital spheres of influence' around the world through exports, investment and development of digital infrastructure (Hemmings, 2020). This will also give them access to new sources of data (digital gold). Access to and control over data is a critical area of competition. The ability to gather, analyse, and utilise data is seen as essential for economic growth and AI development. Yet, perhaps unintentionally, the effect of this is that countries around the world face increasing pressure to align with either the US or China. Some seek to balance their relationships with both, while others are leaning towards one side or the other.
- *Escalation of the digital Cold War*: The US has deepened partnerships with key allies like Japan, South Korea, the EU, and Taiwan to secure semiconductor supply chains and limit China's access to advanced chips. This shows how global tech competition is going beyond the traditional boundary lines of East and West, which adds to the tension. For example, the CHIP 4 Alliance (the US, Taiwan, South Korea and Japan) aims to reduce reliance on Chinese semiconductor manufacturing and coordinate export controls on critical technology. China is strengthening its own tech alliances with Russia, Iran, some Arab States, and BRI nations. Huawei, Alibaba, and Baidu are expanding partnerships in Southeast Asia, Africa, and the Middle East, providing an alternative digital ecosystem outside of US influence. The US and EU are increasing subsidies to their tech industries (e.g., the US$280 billion CHIPS and Science Act) to counter China's state-backed tech giants. The US has pressured allies (Netherlands and Japan) to restrict the export of advanced semiconductor equipment to China. As a result of such US influence, ASML (Netherlands) and Nikon (Japan) now have limited sales of high-end lithography machines to China. China's Digital Silk Road continues to finance and export its tech infrastructure to emerging markets, challenging Western dominance. The US is fostering AI

and quantum computing partnerships with Europe and Indo-Pacific allies to stay ahead of China's advancements. NATO and AUKUS (Australia, UK and US) have incorporated AI and cyber warfare strategies to counter China's growing influence.

- *Digital infrastructure and platforms*: Control over the building of digital infrastructure, including networks, data centres, and undersea cables, is a key battleground. The country that builds the infrastructure is in a position to shape how countries operate in the digital world. The dominance of digital platforms in areas like e-commerce, social media, and mobile payments is a major front in the tech war. The US has historically led in this area, but Chinese platforms are rapidly expanding their influence. Even in the area of military applications, both countries see AI as a key factor in determining military strength and are investing heavily in its development for military use.

The tech war between the US and China is no longer just a bilateral competition—it is a global contest shaped by alliances and partnerships. Both sides are racing to secure supply chains, influence regulations, and dominate emerging technologies, creating a bifurcated digital world where countries are forced to choose between competing tech ecosystems. Beyond the US and China, other nations and regions, particularly the EU, are significantly involved in the tech war between East and West, though often with less direct power than the two main rivals. The EU, though, is not far behind in attempting to gain more sovereignty over digital space, and is at the forefront of democratic societies pushing for more control. The EU is concerned about its reliance on US and Chinese technology, particularly in critical areas like semiconductors and telecommunications. This has led to calls for greater 'technological sovereignty' within Europe. The EU passed a data protection law in 2018 that restricts the transfer of personal data outside of the 27-member bloc and threatens fines to companies that do not protect the sensitive information of EU citizens (Bremmer, 2021). The EU has clamped down on both US and Chinese tech giants through laws like the Digital Markets Act (DMA) and Digital Services Act (DSA) to ensure fair competition. Showing the complexity of the situation, the EU finds itself caught between the US and China, facing pressure to align with US policies while also maintaining strong economic ties with China for manufacturing, electric vehicles (EVs), and green tech. The EU is trying to act collectively to confront tech giants.

It is alarmed, however, that it has not created digital giants the way that the US and China have, and is trying to find new ways to promote its own tech industry. Europe still lacks technology companies with the financial capacity or technological ability to compete against those of the US and China. The EU's push for digital sovereignty is sputtering. The EU's decentralizing trend undermines the possibility of creating a cross-border European telecommunications company. The EU's strategy 'A Europe Fit for the Digital Age,' which was issued in early 2020, aims to improve its digital competition capability (European Commission, 2024). In trying to develop its own tech ecosystem, the EU is actively working to strengthen its own technology sector, investing in research and development, and promoting the development of European tech champions. Its Chips Act (€43 billion investment) seeks to boost Europe's semiconductor production, reducing reliance on TSMC, Intel, and Samsung (Haeck, 2023). The EU is also involved in a trade and technology council with the US to build a 'techno-industrial base' for the US and its partners. The EU shares US concerns about China's human rights record and the security implications of Chinese technology.

In terms of specific EU countries, Germany is a major role player with a large economy and technological capabilities, and is the technological powerhouse of Europe. Despite being more dependent on manufacturing exports than the US, Germany has been slower to respond to the China challenge. The German government has begun to screen Chinese acquisitions of advanced German technology more carefully and launched an EU-wide warning system for critical technologies and infrastructure (Starrs & Germann, 2021). Indicating the EU's multifaceted approach, Germany has also resisted US pressure to preclude Huawei from 5G tenders. The UK is, however, allowing Huawei to build part of its 5G network. The UK also took the lead in joining the China-led AIIB without consulting the US.

In the East, Japan is another major player in the global economy and has a significant technology sector. Japan is working with the US to counter Chinese power in the Indo-Pacific region, specifically focussed on strengthening tech security with the US (Hemmings, 2020). Japan is a crucial ally of the US, but also a key semiconductor and robotics powerhouse with major stakes in the tech war. India, being part of BRICS+aaaaaaaa, is expected to achieve near parity with the US in scientific and human capital and is a growing powerhouse in the global technology sector. It is emerging as a third power, positioning itself as a neutral tech hub that benefits from the US-China rivalry and reducing reliance on both superpowers. As much as it is trying to attract investment from China, India is encouraging Apple, Google, and Microsoft to shift manufacturing away from China and into India.

Balancing between the US and China, South Korea is a key player in the technology sector with companies such as Samsung and SK Hynix. It is continually strengthening cooperation with like-minded partners such as Japan, the US, and EU, to address shared challenges in the technology sector. The South Korean government pledged US$19 billion in AI, quantum computing, and cybersecurity R&D (Dong-Yeon, 2025). Despite US pressure, South Korea still exports some chips to China, fearing economic retaliation and avoiding trade conflicts with Beijing. Again, this shows the complexity and multifaceted nature of the tech war in and between East and West. Taiwan plays a significant role in China's digital development, both economically and culturally. Taiwan is, of course, the world's leading manufacturer of semiconductors. It sees China as a major threat, hence its strong alliance with the US and EU.

Russia seeks dominance in its 'near abroad', meaning it has an ambition to exert control and influence over the countries and regions that border it, which Russia considers to be within its sphere of influence. This includes countries in Eastern Europe, the Caucasus, and Central Asia. Russia views them as vital to its security and geopolitical ambitions as it seeks to control key energy nodes and re-establish its sphere of influence in its near abroad. Russia is also targeting the aggregation of states and alliances that make up Euro-Atlantic power. It views the Western alliance system, particularly NATO and the EU, as a threat to its interests and security. Russia sees the expansion of these Western institutions as an encroachment on its sphere of influence and a challenge to its power. Hence, in this tech war, Russia uses disinformation campaigns, cyberattacks, and the cultivation of political allies to divide, destabilize, and deceive democratic societies (Polyakova & Boyer, 2018). Multiple layers of proxies are used to maintain plausible deniability. Trying to divide the world into blocs and spheres of influence, Russia's strategy is aimed at weakening Western institutions and undermining trans-Atlantic consensus. Russia is also increasingly working with China in opposition to the West, often using nefarious tech methods.

Lastly, countries in Africa, Latin America, and the Middle East are seen as key spheres of influence (perhaps even new 'battle grounds') in the digital competition between the US and China. The US wants to promote Western tech standards in cybersecurity, data privacy, and AI ethics in these regions. China seeks to expand the Digital Silk Road to build 5G networks, smart cities, and AI systems, and provide more affordable tech alternatives (Huawei, Alibaba, and Tencent) to US companies. At least, countries in Africa, Latin America, and the Middle East are benefiting from both sides by securing investments, infrastructure, and digital partnerships. The global tech war isn't just US versus China—it is a complex struggle where Europe, Japan, South Korea, Taiwan, India, and the Global South are all playing crucial roles. In addition, the tech war is not just a competition between countries, but also between corporations. Many tech companies are deeply intertwined across national borders, and it is difficult to disentangle them completely. The multipolar tech war has a significant impact on the global economy, with some companies facing sanctions and trade barriers. It accelerates technological innovation while creating risks of fragmentation in global standards and supply chains. The developments in global tech competition is quite intriguing. While many countries and regions are impacted by the tech war between the US and China, they are also trying to find ways to navigate the situation to their own benefit. Some are attempting to remain neutral, while others are aligning themselves with one side or the other.

Notably, the tech competition between the US and China is not a zero-sum game with a clear winner. Instead, it is a dynamic and complex rivalry characterized by both competition and collaboration, and shaped by state and corporate interests. The competitive landscape is constantly shifting, with China rapidly catching up in many areas, while the US tries to maintain its technological edge. Despite the competition, there is also significant interdependence between the US and Chinese tech sectors, with intertwined supply chains, R&D partnerships, and cross-border investments. While US tech companies have historically held a lead, Chinese tech companies are rapidly closing the gap, especially in areas like FinTech and AI application. The competition is complex, with both sides having strengths and weaknesses. The role of each country's government and national interest remains a major factor shaping the dynamics of this ongoing rivalry.

4 The Digital Revolution: Impacting the SDGs

The digital revolution significantly affects the attainment of the United Nations Sustainable Development Goals (SDGs) by offering transformative opportunities while also posing unique challenges. The digital revolution refers to the shift from analogue and mechanical systems to digital technology, transforming industries, economies, and societies (Dixon, 2023). It began with the rise of computers and the internet in the late twentieth century and continues with advancements in AI, big data, automation, and connectivity. This revolution is reshaping communication, business, governance, and daily life, driving innovation while also raising challenges related to privacy, cybersecurity, and digital inequality. Central to this development is the Fourth Industrial Revolution (4IR), which is characterized by a fusion of technologies that blur the lines between the physical, digital, and biological spheres (Schwab, 2017). It is a new phase in human evolution, facilitated by rapid technological advancements. Digital technologies are at the core of this revolution. The 4IR is different from prior industrial revolutions because of its velocity, scope and impact. Driven by advancements in AI, IoT, blockchain, robotics, and biotechnology, the 4IR is transforming industries, economies, and societies. This digital revolution plays a pivotal role in accelerating progress toward the SDGs, but also presents new risks and challenges that could hinder equitable development.

Before considering these risks and opportunities, it is important to note that the 4IR is quite unique in enabling the use of advanced digital technologies (also known as Industry 4.0) to create more efficient, sustainable, and inclusive solutions across various sectors, and contributing to progress on goals related to poverty reduction, healthcare, education, climate action, and more. The convergence of these trends highlights key areas of synergy and impact (Alhammadi et al., 2024). As part of the 4IR, Industry 4.0 technologies are, for example, driving sustainable manufacturing, minimizing environmental harm, and promoting responsible production, aligning with SDG 12 (Responsible consumption and production). Another example is their integration into urban development, which is facilitating the growth of smart cities, supporting SDG 11 (Sustainable cities and communities). Furthermore, Industry 4.0 plays a vital role in advancing clean energy solutions, contributing to SDG 7 (Affordable and clean energy) and SDG 13 (Climate action). Additionally, the application of data analytics within Industry 4.0 enhances informed decision-making, which is essential for achieving multiple SDGs, including SDG 4 (Quality education) and SDG 8 (Decent work and economic growth). In the broader context, there are a number of ways in which the 4IR is influencing the impact of digital technologies on countries' ability to achieve the SDGs, such as:

- *Being a catalyst for digitalization*: The 4IR is driving the rapid development and integration of digital technologies. This is leading to a more interconnected and digitized world, with digital technologies playing a central role in efforts to achieve the SDGs.

- *Industry 4.0 and digital transformation*: The term Industry 4.0 was coined to address how the 4IR relies on digitalization rather than automation (Mhlanga, 2021). Industry 4.0 is characterized by the automation and digitization of production, human–machine interaction and automatic data exchange. Digital transformation, driven by 4IR technologies, impacts most of the SDGs.
- *Specific technologies and precision solutions*: The 4IR is characterized by the emergence of technological breakthroughs in AI, advanced robotics, neurotechnology, synthetic biology, 3D printing (additive manufacturing), cloud and quantum computing, 5G and energy storage, among many others (Bai et al., 2023). These technologies are being applied in various sectors, including health, agriculture, and manufacturing to support the achievement of the SDGs.
- *Digital financial inclusion*: FinTech and 4IR technologies such as AI, big data, and blockchain are being used to advance digital financial inclusion by increasing access to financial services, reducing costs, and improving efficiency (Mhlanga, 2022). This can play a significant role in reducing poverty and promoting economic growth.
- *Data and knowledge*: Big data and digital technologies are shaping knowledge for sustainable development and the SDGs. The 4IR is creating unprecedented amounts of data, which can be used to track progress on the SDGs, in order to identify areas where more action is needed.
- *Smart technologies*: Smart technologies, digitalization, and data are central to urban development and sustainable cities (Gupta et al., 2023a). These technologies are used to leverage innovative infrastructure, such as intelligent transport systems and smart water management. Smart technologies such as AI-driven predictive analytics, IoT-enabled sensors (e.g., for more effective waste disposal and optimized energy use), and smart grids play a vital role in enhancing efficiency, reducing environmental impact, and improving the quality of life for residents.
- *Inclusive innovation*: 4IR technologies can empower marginalized communities by providing access to information, education, and financial services, contributing to reduced inequalities (SDG 10) and inclusive and sustainable economic growth (SDG 8).

Initially starting with the Millennium Development Goals (MDGs) in 2000 and then the SDGs in 2015, the United Nations set out to create a guideline for nations to fast-track economic development to achieve them by 2030. Originally, the MDGs was introduced as an eight-goal agenda to address extreme poverty, hunger, disease (health), lack of education, gender inequality, environmental degradation, and to promote global partnerships—in order to drive economic and social progress (United Nations, 2015). The primary intention was to mobilize global efforts toward measurable improvements in key areas of human development, especially in the Global South. Building on the MDGs' progress, the SDGs were adopted with a broader and more ambitious framework. The 17 SDGs aim to address the root causes of poverty and inequality, while also integrating environmental sustainability, economic development, and peace-building (UNDP, 2025).

Unlike the MDGs, the SDGs apply to all countries, recognizing that sustainable development is a global responsibility. In short, while the MDGs focused on urgent human development challenges, the SDGs expanded the vision to long-term sustainability, equity, and global cooperation to create a better future for all.

Given that the digital revolution worldwide has contributed significantly to how the world has changed and to the nature of human progress (i.e., innovation), and specifically new economic opportunities, the following question arises: In what ways are the digital revolution influencing progress toward achieving the SDGs? It is clear that this revolution plays a pivotal role in accelerating progress toward the SDGs, but also presents new risks and challenges that could deter equitable development. On the positive side, and given the rapid advancement and integration of digital technologies in society, it may present the following opportunities (linked to the SDGs):

- *Acceleration of progress*: Digital technologies can act as a catalyst, accelerating progress across all 17 SDGs. The digital revolution has the potential to transform infrastructure, health and education systems, and resource consumption. Digital paradigms are considered enablers of sustainable development, as they can support human progress, enhance personalization, and reframe the business case for both investors and consumers, transforming behaviours and enabling a circular economy (Xiong & Dai, 2023). In this regard, the digital revolution specifically helps countries to achieve SDG 8, SDG 9 (Industry, innovation and infrastructure), SDG 11, SDG 12, and SDG 13.
- *Enhanced connectivity*: Improved access to information and communication technologies (ICTs) can connect people, facilitate knowledge sharing, and empower communities. Digital technologies connect people to financial, social, and economic systems (e.g., mobile banking and blockchain-based remittances). This can improve access to education, healthcare, and financial services, contributing to goals like poverty reduction (SDG 1) and quality education (SDG 4).
- *Data revolution*: Digitalization offers a 'data revolution' for the SDGs, providing new sources of data, enhanced analytical capacities, and collaborative digital ecosystems. Novel data sources, such as earth observation data and mobile data, can be used to track progress on SDG indicators. Specifically, the use of AI and 'big data' can help to attain SDG 13 through climate modelling, carbon footprint tracking, and smart energy solutions to support sustainability efforts (Shahzad et al., 2022). Digital paradigms and ICTs are essential to overcome critical data gaps, by complementing traditional data sources and enhancing granularity.
- *Financial inclusion*: FinTech and related technologies are key enablers of financial inclusion. They can provide access to financial services, especially for those who are unbanked, and promote economic empowerment (Mpofu, 2023). Digital financial services (e.g., mobile banking, digital wallets, decentralized finance (DeFi), and blockchain-based remittances) are increasing financial inclusion, especially in underserved regions, creating the opportunities to reduce poverty

(SDG 1) and closing the digital divide to reduce inequalities (SDG 10). Financial inclusion fosters entrepreneurship and job creation (SDG 8) by providing access to credit and other financial services for small and medium-sized enterprises (SMEs). Financial inclusion can also facilitate investments in education (SDG 4). For instance, Islamic digital finance are facilitating investments in education, including funding for schools, universities and educational programs that foster skill development (Tlemsani et al., 2023). Financial inclusion can improve access to healthcare by enabling individuals to afford medical expenses and purchase health insurance (SDG 3).

- *Digital social innovation*: Digital technologies and social innovations are combining to create new solutions for social problems at scale. Digital social innovations connect the innovation process, the social world, and the digital ecosystem. They are being applied to healthcare, smart cities, agriculture, and poverty reduction. For example, crowdsourced food redistribution (e.g., OLIO and Too Good To Go) are digital platforms that can connect surplus food from businesses and households to people in need, reducing food waste and addressing hunger (SDG 2). Crowdfunding platforms can enable individuals and communities to raise funds for various social causes, empowering them to address local challenges and support community development initiatives (SDG 8). E-governance can also play a vital role in achieving SDG 16 (Khan et al., 2022). Digital platforms can improve government efficiency, transparency, and accountability by facilitating online service delivery, citizen engagement, and access to information. E-governance helps to reduce the rural–urban divide through access to health, education, and the rising non-agricultural income-generating opportunities for rural areas. Big data and AI platforms, such as DataKind and Microsoft AI for Good, can be used for humanitarian aid to improve disaster response, refugee assistance, and social development initiatives.
- *Smart and sustainable cities*: Digital technologies are essential for the development of smart and sustainable cities and communities (SDG 11). They are used to improve transportation, water management, and energy efficiency, resource management, among other things. Digital technologies are crucial for the development and deployment of renewable energy sources, such as solar and wind power, contributing to climate action (SDG 13) and affordable and clean energy (SDG 7). The implementation of IoT devices that provide integrated monitoring of physical spaces improves security and response times (Dionisio et al., 2024). In addition, smart mobility and Society 5.0[1] are high-potential concepts

[1] Society 5.0 is a vision for a human-centered, super-smart society that integrates advanced digital technologies such as AI, IoT, big data, robotics, and blockchain to solve societal challenges and improve quality of life. The evolution of societies can be summarized as: Society 1.0 (hunting and gathering), Society 2.0 (agricultural society), Society 3.0 (industrial society), Society 4.0 (information society), and Society 5.0 (super-smart society). The latter envisions a future where humans and technology coexist harmoniously, solving global challenges through digital innovation while ensuring economic, social, and environmental sustainability.

for promoting sustainable cities, born from disruptive technologies. Barcelona's smart traffic lights is an example, where it uses IoT sensors and AI to optimize traffic flow, reducing congestion and emissions. Another example is Seoul's smart bins, where IoT-enabled bins monitor waste levels and optimize collection routes, reducing environmental impact.

- *Improved healthcare*: e-Health technologies like telemedicine (e.g., Babylon Health and TeleDoc), AI-powered medical diagnostics (e.g., Google's DeepMind for eye disease detection), IoT, and personalized medicine can improve healthcare access and quality, contributing to good health and well-being (SDG 3). AI and machine learning contributed significantly to managing the COVID-19 pandemic globally (Mhlanga, 2022). AI models help detect diseases such as diabetic retinopathy, improving early diagnosis and treatment. Digital health records and predictive analytics improve healthcare efficiency and disease prevention. In terms of IoT, the use of sensors in both patients and operating rooms provide physicians with real-time support during procedures. Healthcare can be personalized through remote healthcare consultations and wearable health-monitoring devices. This would reduce the strain on hospitals and improve healthcare access in underserved areas (SDG 10).
- *Precision agriculture*: Digital technologies like AI and IoT are being used in agriculture to optimize practices, improve yields, and enhance sustainability (e.g., minimizing environmental impact). This can contribute to achieving food security, zero hunger (SDG 2) and sustainable agriculture (SDG 15). In India (as in other parts of the world) for instance, farmers do AI-powered crop monitoring, receiving real-time data on soil quality, crop health, weather patterns, irrigation optimization, and pest infestations, which improves yields and reduce food waste (Tlemsani et al., 2023). Smart agriculture also involves supply chain tracking and digital marketplaces, which help to improve food security and reduce food waste.
- *Monitoring and awareness*: Digital technologies facilitate the monitoring of progress toward SDGs and raise awareness about sustainability issues. In, for instance, trying to achieve SDG 14 (Life below water) and SDG 15 (Life on land), the predominant contribution of disruptive technology will be monitoring and promoting awareness about current progress (Dionisio et al., 2024). Put differently, accurate weather forecasting and real-time monitoring of ecosystems (SDG 13) would not be possible without the aid of disruptive technologies (e.g., IBM's Environmental Intelligence Suite for air quality monitoring and predicting climate risks, using AI and data analytics). ICT is also crucial in fostering multi-stakeholder alliances to enhance data monitoring and accountability (SDG 17—Partnerships for the goals). Another example is how platforms of electronic participation are using monitoring technologies, interactive apps, and social media to implement citizen observatories to prevent flood risk management and aiding in disaster preparedness and response, with citizens, public and private sectors working together (Vyas-Doorgapersad, 2022). This helps to achieve a number of SDGs: SDG 11 (Sustainable cities and communities), SDG 13, SDG 6 (Clean water and sanitation), and SDG 15.

- *Industry 4.0 and 4IR*: Digital technologies are central to Industry 4.0, which is driving innovation in manufacturing and other industrial sectors and creating opportunities for sustainable economic growth. Disruptive technologies underpin the 4IR transition to Industry 4.0 and Industry 5.0, enhancing productivity, efficiency, connectivity, resiliency, and sustainability (Khan et al., 2022). Smart technologies, digitalization, and data are imperatives in achieving SDG 9. Industry 4.0 strongly focus on digital transformation in manufacturing and business, increasing productivity and efficiency. The term Industry 4.0 was coined to address the fact that the 4IR will rely on digitalization rather than only automation as the third did, and that future production will be modular within factories consisting of 'smart' objects (Vyas-Doorgapersad, 2022). Taiwan, in particular, is making gigantic advancements in Industry 4.0, focusing on automation, robotics, and IoT technologies, with companies providing sensors, industrial PCs, and automation solutions. Industry 4.0 technologies can have a beneficial impact on all 17 SDGs. This is even more so with Industry 5.0, which focuses on human-AI collaboration, sustainability, ethics, and resilience while leveraging Industry 4.0 technologies (Mondejar et al., 2021). In line with the SDGs, Industry 5.0 is a response to challenges like climate change (SDG 13), economic inequality (SDG 10), and the need for ethical AI governance (SDG 16). Unlike Industry 4.0's profit-driven model, Industry 5.0 balances efficiency with social impact (SDG 12), ethical AI, and sustainability (Kasinathan et al., 2022). It also strengthens supply chain resilience against global disruptions. This evolution in the 4IR brings a shift from just smarter factories to more sustainable, human-centred, and adaptable industries. Table 2 shows a comparative summary with regards to evolution in the 4IR.

The digital revolution also presents challenges related to inequality, security, and ethical concerns. These and other factors are having a negative impact on countries' attainment of the SDGs—often due to the tension between profit-driven productivity (mainly by corporate enterprises) and socio-environmental factors (mostly dealt with by governments). The following challenges and risks can be identified in this regard:

- *Digital divide*: Unequal access to technology and digital literacy can exacerbate existing inequalities, hindering progress on SDG 10 (reduced inequalities), which could also have other negative impacts on the environment, society, and individuals. Access to technology is uneven across regions and sectors, deepening socioeconomic disparities. For instance, the digitization of finance might lead to the exclusion of vulnerable groups, such as the poor, the illiterate, and the elderly as financing decisions and transactions are automated (Mpofu, 2023). Progress in harnessing digital paradigms for pursuing the SDGs is very slow-paced, incremental, and highly context-dependent, and some SDG gaps rely on structural, cultural, financial or human factors, which are not susceptible to being solved by technological innovations (Sparviero & Ragnedda, 2021). The digital divide, which has become the new face of inequality, remains a

Table 2 Comparing Industry 4.0, Industry 5.0 and Society 5.0

Aspect	Industry 4.0	Industry 5.0	Society 5.0
Focus	Automation, IoT-efficiency, digital transformation, and smart manufacturing	Human-centred integration with AI and robotics, and resilient industrial systems	Integrating digital technology into society for sustainability and inclusivity
Key technologies	IoT, AI, big data, robotics, cloud computing, cyber-physical systems, and automation	Human-AI collaboration, virtual reality, personalized automation, collaborative robots (cobots), ethical AI	AI, 5G, robotics, biotech, blockchain, IoT, digital twins (metaverse), smart infrastruc-ture and quantum computing
Role of humans	Reducing human involvement through automation and robotics	Reintegrating human intelligence alongside AI and machines for creativity and customization	Humans are the priority—technology is designed to enhance societal well-being
Goal	To optimize production and productivity through data-driven decision-making and minimal human intervention, thus increasing efficiency and automation in manufacturing	To harmonize human and machine collaboration, emphasizing creativity and well-being; ensuring that technology serves society, focusing on personalization, resilience and sustainability	To balance economic progress with social well-being, using digital innovation, and to solve social issues (e.g., poverty) by integrating digital tech into all aspects of life
Customization	Mass production with standardization, automation and efficiency as priorities	Mass personalization and customization using human- centric AI-driven insights	Personalized digital services for diverse societal needs that adapt to individual needs
Sustainability	Very efficiency-driven but limited environmental and social impact considerations	Strong emphasis on green manufacturing, circular economy, ethical AI, and reducing carbon footprints	Holistic sustainability—social, environmental, and economic balance
Economic model	Profit-driven—maximizing production efficiency (mass and quality through precision tech), reducing costs, and optimization	Value-driven—balancing profit, social impact, well-being, and sustainability. Mass personalization and customization	Balance-driven—socio-economic balance (inclusive economy)—technology serving both economic growth and public welfare
Resilience	Vulnerable to global supply chain disruptions, cyber threats, and system failures (technology dependence)	More adaptable—builds resilience into supply chains and industrial processes to withstand various crises	Highly resilient—integrates technology for crisis response, adaptability, and inclusive growth
Ethical considerations	AI-driven decision-making without much ethical oversight. Concerns about data security, privacy, and job displacement	Ethical AI, data privacy, socially responsible innovation, fair labour practices, and human-centred design	Responsible AI, inclusive technological access, and balancing technological progress with human rights, inclusivity and sustainability

Sources Dionisio et al., (2024), Mondejar et al. (2021), Bai et al., (2023), Richnák and Fidlerová (2022) barrier and is egregious to the ethos and the reality of achieving the SDGs, especially given that more than four billion people lack access to the internet in emerging economies. While connectivity can reduce the digital divide, it is meaningless without proper digital literacy. The ripple effects of this is that it would certainly widen not only the gap between developed and developing nations, but also the gaps within nations. Hence, a fundamental concern for the worsening of economic inequality, according to Mirrlees (2024), is that the global 'tech war' is a class conflict, benefiting the bourgeoisie at the expense of workers in the countries that are at 'war'.

- *Environmental impact*: Data centres and the production of electronic devices have significant environmental impacts, potentially undermining efforts to combat climate change (SDG 13). E-waste is generated as rapid tech advancements (frequent upgrades) lead to short device lifespans, which negatively affect SDG 12 (Responsible consumption and production). It often results in growing mountains of e-waste, which contains hazardous materials like heavy metals and toxic chemicals (Mondejar et al., 2021). Making it worse, much e-waste is improperly disposed of, often ending up in landfills or being shipped to developing countries where it is processed in unsafe and environmentally damaging ways, causing pollution and affecting biodiversity. This poses risks to human health and ecosystems (SDGs 3, 11, 12, 15). High energy consumption is another concern as data centres, cryptocurrency mining, and AI models require vast amounts of electricity, contributing to greenhouse gas emissions (affecting SDG 7 (Affordable and clean energy) and SDG 13). This is worsened by the production and use of billions of electronic devices, from smartphones to laptops, which require significant energy input (Mpofu, 2023). Resource depletion is also causing distress as digital devices rely on rare-earth minerals—like europium for LEC screens, dysprosium for electric vehicles, and neodymium for hard drives—leading to unsustainable mining practices (affecting SDG 15). Water usage for data centres is another form of resource depletion as cooling systems consume large amounts of water, contributing to water stress (affecting SDG 6). Ecological concerns are further exacerbated by the growing digital carbon footprint as these data centres, the internet (including routers, servers, and cables), cloud computing, and 5G networks generate CO_2 emissions (affecting SDG 13). Lastly, the expansion of data centres and infrastructure through green investments in digital technologies might indirectly fuel deforestation, unsafe energy usage, and habitat loss (affecting SDG 15).
- *Job displacement*: Automation and AI could lead to job displacement in traditional industries, impacting decent work and economic growth (SDG 8). Concerns exist about the effects on the labour force and the possible instability and impact it may cause in low-skill/low-pay jobs (Dionisio et al., 2024). It is expected to increase unemployment, while adequate reskilling efforts may be limited or less successful due to age, lack of willingness and other demographic factors. In addition, job replacement may actually occur faster than reskilling efforts can keep up.

- *Data privacy, cybersecurity threats and misinformation*: Cases of fraud, robberies, sexual harassment of women, gender discrimination, and violence against women are perpetuated (Mpofu, 2023). These concerns around data privacy, security, and misuse of information particularly have a negative effect on achieving SDGs 1, 5 (Gender equality), 8, and 10. Cyberattacks can disrupt critical infrastructure, hinder development efforts, and undermine progress on various SDGs. Specifically, SDG 16 (Peace, justice, and strong institutions) may be most negatively affected as AI-driven misinformation, cyber threats, and data privacy concerns (e.g., risks associated with FinTech) destabilize institutions. AI-driven surveillance could also undermine democracy and privacy rights.
- *Ethical and social concerns*: Digital technologies are changing the information landscape, raising ethical and social concerns. The digital revolution presents regulatory challenges, which is negatively affecting SDG 17 as it places global cooperation under strain, raising the need for overly demanding collaboration to ensure ethical AI development and digital governance. There is also the risk of AI and algorithmic bias and discrimination (Pigola et al., 2021). AI systems can inherit biases from the data they are trained on, which can perpetuate and amplify existing discrimination (e.g. gender bias), particularly in areas like hiring, law enforcement, and lending (credit scoring). In this regard, SDGs 5 and 8, in particular, are at risk of being negatively affected. In addition, those without access to digital technologies may be excluded from essential services, information, and economic opportunities, which is negatively affecting SDGs 1, 2, 3, 5, 8, 10, and 16. Likewise, digital financial illiteracy can lead to increased risks and fraud and can increase inequalities. The ethical use of public data is another concern, for ethical issues arise with the increased use of public data. It is important to note, though, that there is a lack of consensus about ethics, responsibility, or 'social good' in the context of digitalization (Mondejar et al., 2021).

On the last issue, addressing these ethical concerns requires a balanced approach to policy-making, regulation, and corporate responsibility to ensure that the benefits of the digital revolution are accessible to all while minimizing its negative impacts (Kasinathan et al., 2022). By proactively attending to these ethical challenges to pave the way towards SDG-attainment by 2030, we can ensure that the digital revolution serves humanity and contributes to a more just and equitable world. It requires a multi-faceted approach, including:

- *Data protection regulations*: Implementing strong data protection regulations to safeguard personal information.
- *Ethical AI development*: Developing ethical guidelines and standards for the development and use of AI.
- *Bridging the digital divide*: Investing in digital infrastructure and digital literacy programs to ensure equitable access to technology.

- *Combating misinformation*: Promoting media literacy and fact-checking initiatives to combat the spread of misinformation and disinformation.
- *Promoting digital rights*: Protecting fundamental rights such as privacy, intellectual property, freedom of expression, and access to information in the digital age. The ethical use of technologies to protect human rights should be a collective global exploration with a genuine intention to find a healthy balance between protection and freedom.

This will require building a sensible and impartial techno-ethical framework, involving: (1) ethical governance to implement policies that ensure the fair and safe integration of digital technologies; (2) promoting digital literacy, that is, fostering digital literacy, innovation, and investments to drive Agenda 2030; (3) addressing social dilemmas through further research on ambivalent impacts, covering all relevant contexts and communities, particularly in the Global South, to reduce digital divides; (4) fair data and transparency that encourages accountability and transparency to combat discrimination and biased outcomes; (5) multi-stakeholder alliances, meaning alliances for social good to bring relevant stakeholders together, including civil society and vulnerable communities, to share data and overcome current capacity and accessibility constraints; (6) integrating ethics into AI to ensure the proper use of AI for the 2030 Agenda; and (7) just digital guidelines that prioritize fair access to digital capabilities, commodities, infrastructure, and governance to reduce global inequality and advance the SDGs.

Possibly the best available example (or starting point) of such a techno-ethical framework to speed up digital transformation[2] at the country level in an inclusive way, is that assembled by the UNDP (2022) in its 10 illustrative practices that governments can consider. It shows that digital transformation must be intentionally inclusive if it is to contribute to realizing the SDGs. Such an inclusive approach involves the whole-of-society as well as the development of "holistic digital ecosystems that leave no one behind" (Zakhozhyi, 2023: 1). This entails uniting all sectors of society—governments, the private sector, civil society, and citizens—to build or enhance local digital ecosystems. An intentionally inclusive approach ensures that individuals, governments, non-profits, and businesses alike can reap the benefits. When purposefully planned and implemented at the national level, digital transformation empowers countries to embrace and utilize digital technologies effectively, systematically, and sustainably, fostering social cohesion rather than deepening divides. To foster inclusive digital transformation at the national level, governments can draw inspiration from the 10 illustrative practices outlined in Table 3. These best practices or action steps are derived

---

[2] Digital transformation is the integration of digital technologies into all aspects of society, business, and governance, changing how economic and social activities are conducted to improve efficiency, accessibility, and innovation. It leverages 4IR technologies to reshape operations, services, and interactions. Beyond just adopting technology, it requires cultural and structural shifts to ensure sustainable, inclusive, and effective use of digital solutions.

from the UNDP's experiences, insights, and lessons learned while supporting digital transformation efforts in over 100 countries over the past six years. They can be implemented randomly, meaning governments can start with any of the action steps, proceeding in any order or preference, as per their needs and circumstances. There is no specific (or Taylor-made) sequence, but implementing all of the practices is recommended.

It is essential to note that accelerating ethical and inclusive national digital transformation requires a range of targeted interventions and a comprehensive, whole-of-society approach. The UNDP's extensive experience with countries demonstrates that, when applied effectively, the recommended good practices can pave the way for successful digital transformation efforts. Following a holistic approach will also help address ethical challenges like data privacy and security, digital inclusion vs. digital exclusion, surveillance and monitoring, digital sovereignty, and others more effectively. Successfully managing these ethical challenges through responsible policies, governance, and best practices is crucial for ensuring a truly sustainable and equitable digital transformation. In line with Agenda 2030, the vision of Society 5.0 is that of a super-smart society where digital innovation enhances overall well-being. This will ensure that economic growth is aligned with social and environmental sustainability. However, considering the traditionally high social and ecological costs of conventional economic growth and people—being creatures of habit—often treating digital change negatively, this could be more of a challenge than anticipated. That is why the SDGs play such a vital role in shaping shared objectives for the whole of humanity, creating valuable synergy between human needs and social innovations that is human-centric.

While the digital revolution presents transformative opportunities for achieving the SDGs, it also requires strong governance, ethical AI practices, and global cooperation to ensure that technological advancements contribute to inclusive and sustainable development. The 4IR has the potential to drive unprecedented progress toward the SDGs, but without inclusive policies and global cooperation, it could exacerbate inequalities and security risks (Mpofu, 2023). As key stakeholders, governments, businesses, and civil society must work together to harness 4IR technologies responsibly for sustainable and equitable development. Overall, the 4IR presents both opportunities and challenges for SDG attainment. Digital technologies, driven by the 4IR, have the potential to advance progress on the SDGs substantially, but these technologies must be developed and implemented responsibly, ethically, and inclusively to ensure that their benefits are widely shared and their potential harms mitigated.

To maximize the potential of the digital revolution for SDG achievement, it is crucial to implement inclusive digital policies, harness investments in digital infrastructure and human capital, and focus on ethical and responsible technology development. To ensure that the digital revolution works for the SDGs, the following leapfrogs are required:

Table 3 Framework for digital transformation

Practices (action steps)	Description and relevance to the techno-ethical framework	Contribution to realizing the SDGs
Practice 1: Introduce participatory policy-making and governance for inclusive digital transformation	Designing inclusive policies and digital transformation processes requires diverse perspectives. Policy-making should be grounded in evidence, oversight should be transparent, and vulnerable groups must have representation at every level	It ensures that technology-driven progress benefits all segments of society. It promotes fairness, social cohesion, and sustainable develop-ment. SDGs positively impacted: SDG 1, 4, 5, 8, 9, 10, 16 and 17
Practice 2: Strengthen institutions to lead digital transformation	Effective institutional governance mechanisms are essential for achieving inclusive digital transformation, particularly in upholding rights. Government institutions must be well equipped to fulfil their roles, and it is crucial to design or adapt them to align with the digital era	Well-structured institutions help create transparent policies, protect human rights, and promote equitable access to digital opportunities. By adapting governance frameworks to the digital era, states can enhance public service delivery, reduce corruption, and drive social and economic progress. SDGs positively impacted: SDG 9, 10, 16 and 17
Practice 3: Strengthen national digital safeguards, laws and standards	Digital transformation requires a rights-based regulatory and legal evolution, as regulations translate principles into enforceable regulations. Such an approach leverages digital transformation to expand freedoms and promote inclusivity. Additionally, specific safeguards must be implemented to mitigate the risks associated with digital technologies	Strong digital regulations create a safe and equitable digital environment, fostering sustainable development and social justice. It protects freedoms, and mitigates risks such as discrimination, surveillance, and data misuse. SDGs positively impacted: SDG 5, 9, 10, 16 and 17

(continued)

Table 3 (continued)

Practices (action steps)	Description and relevance to the techno-ethical framework	Contribution to realizing the SDGs
Practice 4: Invest in responsible data collection and usage	For inclusive digital transformation, countries must enhance rights-based data collection. Regularly tracking digital inclusion metrics—like internet access, mobile subscriptions, and digital literacy—supports informed, unbiased decision-making. Disaggregated data helps tailor policies to people's needs	Rights-based data collection strengthens digital policies by ensuring inclusivity, reducing biases, and enabling evidence-based decision-making. This helps states design targeted interventions, ensuring no one is left behind in the digital transition. SDGs positively impacted: SDG 1, 4, 5, 9, 10 and 16
Practice 5: Invest in meaningful universal connectivity	Globally, 2.7 billion people lack internet access. Meaningful connectivity goes beyond availability—it's broadband that is accessible, affordable, relevant, safe, trusted, and empowering, creating positive impacts. Connectivity is key to reducing the digital divide, yet connectivity alone is insufficient without digital literacy. The latter should as much be prioritized, both by private sector investment and public sector capacity-building and incentives	Connectivity is vital for bridging the digital divide and ensuring that all individuals, regardless of their location or status, have access to opportunities in education, jobs, and healthcare. Combining connectivity with digital literacy ensures that individuals can effectively engage with digital tools, creating lasting positive outcomes. SDGs positively impacted: SDG 1, 4, 5, 9, 10 and 16
Practice 6: Advocate for appropriate use of digital public goods (DPGs)	DPGs are open-source, allowing countries to study, adapt, and implement them quickly rather than starting from scratch. Using DPGs enhances digital sovereignty and avoids vendor lock-in, as countries control the software and data. Sectoral DPGs, such as those for digital ID, payments, or data exchange, can combine to form key digital public infrastructure	DPGs promote digital sovereignty and empower countries to build their own digital infrastructure, which can enhance inclusion, reduce dependencies, and foster sustainable development. By facilitating open-source solutions, countries can create scalable, secure systems that drive progress across various sectors like education, finance, and governance. SDGs positively impacted: SDG 1, 4, 5, 9, 10, 16 and 17

(continued)

Table 3 (continued)

Practices (action steps)	Description and relevance to the techno-ethical framework	Contribution to realizing the SDGs
Practice 7: Build national digital literacy and capacity	Investing in foundational digital literacy is essential, but not enough. It is also crucial to expand training and employment opportunities in areas like technology, design, research, and agile delivery. Enhancing digital skills and national capacity creates new job and growth prospects. Digital literacy programs must address risks related to data privacy, disinformation, and hate speech. Inclusive digital literacy should focus on marginalized groups and their specific needs. Also, public sector leaders need the skills and confidence to guide national digital transformation for the benefit of all	Investing in digital literacy and skills development is essential for creating equitable opportunities in the digital economy. By enhancing digital capabilities across sectors and among marginalized groups, countries can foster inclusive growth, improve employment prospects, and support informed participation in the digital world. Strengthening national digital capacity also enables governments to lead effective and inclusive digital transformations. SDGs positively impacted: SDG 1, 4, 5, 8, 9, 10, 16 and 17
Practice 8: Build diverse, interdisciplinary teams	Digital transformation can only succeed with empowered teams, not a single coder. Teams developing digital policies, programs, or services benefit from diverse identities and experiences, helping them better understand underserved groups' needs, engage with diverse communities, and anticipate unintended consequences	Empowering diverse teams in digital transformation ensures more inclusive, equitable, and effective policies and services. Different perspectives help identify underserved groups' needs, reduce biases, and create solutions that work for all. This approach fosters social cohesion, addresses inequalities, and strengthens digital initiatives' impact. SDGs positively impacted: SDG 1, 4, 5, 10, 16 and 17

(continued)

Table 3 (continued)

Practices (action steps)	Description and relevance to the techno-ethical framework	Contribution to realizing the SDGs
Practice 9: Support inclusive business models	Countries can foster and expand inclusive business models by supporting the development of affordable digital products, creating new markets and revenue streams, or incentivizing companies to offer additional services for those with limited digital literacy	By promoting inclusive business models, countries can drive economic growth, enhance digital access, and reduce inequalities. Supporting affordable digital products and services for people with limited digital literacy helps bridge the digital divide, ensuring that all are included in the transformation process. SDGs positively impacted: SDG 1, 4, 5, 8, 9 and 10
Practice 10: Promote inclusive service design	Intentionally inclusive service design addresses the needs of a wider range of people. Effective services must solve a clear problem while being accessible and usable by everyone, regardless of their circumstances or abilities	Such a design ensures that solutions are accessible to all, including marginalized and vulnerable populations. By addressing diverse needs, such services improve equity, enhance well-being, and foster social inclusion, contributing to broader sustainable development goals. SDGs positively impacted: SDG 1, 4, 5, 8, 10 and 16

Sources UNDP (2022), Zakhozhyi (2023)

- Sound enabling frameworks and responsible principle-based regulation.
- Funding mechanisms, tailored incentives, and sustainable business models allowing collaboration across stakeholders.
- Coordinating international efforts among all the ecosystem actors by building-up public–private multi-sectorial partnerships and promoting market demand.

5 Conclusion

According to UNDP Administrator, Achim Steiner (2023: 1), "The future is digital, defining how we will live, work, and interact with each other. Whether technology becomes an empowering force for good or a sower of more division and exclusion will depend on the choices we make now". As a force for good, digital transformation can significantly accelerate the implementation of the SDGs. On the other hand, as a 'sower of division', the United States and China are fiercely engaged in a 'digital tech war' that includes industries ranging from computer hardware and software to telecommunications and social media platforms, to AI and digital payment systems. Representing the great powers of the East and the West, the China-US tech war is more than just an economic battle—it is a fight to define the rules of global governance for the digital age.

This technology rivalry has escalated significantly since 2017 through various measures, such as tariffs, export controls, and market access restrictions. This conflict, focused on dominating future technologies like semiconductors and AI, has already resulted in China losing nearly US$150 billion in exports to the US, while reshaping trade flows with the diversification of US imports from countries like Mexico, Taiwan, and Vietnam (Duthoit, 2024). Despite growing tensions, economic cooperation between these two giants is still crucial. Nearly 30% of semiconductor manufacturing machinery exported by the US is still directed to China. For instance, a significant portion of US electronics imports from countries like Vietnam, Taiwan, and Mexico contain Chinese components. This highlights China's continued importance in the global electronics value chain, both as a supplier and a consumer. The same is true in other tech market segments as well. This is the irony of today's technology-driven dichotomy of 'globalization/fragmentation'. The US and China leads the 'tech war', but they also need each other, given growing tech-economic interdependency.

US-China competition is even evident in how President Donald Trump convinced Panamanian President José Raúl Mulino within weeks after his second-term started in January 2025 to end its agreement with China for managing the Panama Canal. An action that China condemned, but the agreement was never to outsource it to another nation like China (Sharp, 2025). The way the US is now gaining control over its strategic interests is certainly changing the global dynamics. Yet, American companies face a dilemma. Despite pressure to lessen their reliance on China, they still view it as a key trading partner. Over the past decade, US companies have secured 54% of the profits in the global electronics industry, a figure that increases to 88% when including their Japanese, South Korean, and

Taiwanese counterparts (Duthoit, 2024). Over the same period, despite growing sales and major technological progress, Chinese companies have only captured 7% of the global electronics industry's profits and lag far behind the leaders in the critical semiconductor sector. Being both a driver and a reflection of the 4IR, the global electronics industry is a good indicator of dynamics in newly emerging technologies as it plays a central role in the development and deployment of many of these innovations.

Specifically, AI plays a pivotal role in the competition between the US and China, becoming a key area of technological and economic rivalry. Both countries are investing heavily in AI research, development, and application, as AI is seen as a critical driver of future economic growth, national security, and global influence (Mirrlees, 2024). AI is becoming a central battlefield for technological and geopolitical dominance. Here are several ways AI is fuelling the tech war:

- *Technological superiority and global influence*: AI is a foundational technology shaping healthcare, transportation, finance, military, and manufacturing. Dominating AI means securing technological leadership and global influence. The US has led in AI innovation, but China is rapidly closing the gap through major investments, partnerships, and a focus on AI-driven industries (Erie & Streinz, 2021). Chinese companies, particularly Huawei, are leading the development of global 5G standards, marking the first time during the internet era that Chinese companies enjoy technological and commercial leadership.
- *Military implications*: AI is regarded as critical for national security, with both China and the US investing heavily in AI for defence technologies like autonomous weapons, surveillance, cyberwarfare, and intelligence gathering. Despite raising ethical concerns, control over AI-driven military technology is viewed as a key factor in maintaining military dominance and influence. AI is used to analyse vast amounts of data from various sources (e.g., satellites, drones, sensors) to enhance situational awareness on the battlefield and improve military decision-making (Mirrlees, 2024). Both the US and China use AI for both offensive and defensive cyber operations, including threat detection, automated incident response, and the development of sophisticated cyber weapons.
- *Economic competition*: AI is seen as a key driver of innovation and economic growth. It can drastically enhance economic productivity by improving efficiencies, creating new products, and optimizing and reshaping global supply chains (Mhlanga, 2022). The US and China both see AI as central to maintaining or increasing their global economic influence. Complicating matters, however, is the fact that both the US and China host the world's largest digital tech sectors, which are interconnected through investments, R&D partnerships, and supply chains.
- *AI in strategic industries*: AI is becoming integral in many key industries, especially:
  - Semiconductors: The US and China are both focusing on developing AI-optimized semiconductor technologies. China aims to reduce its reliance

  on US semiconductor companies, while the US seeks to maintain its leadership (in collaboration with Taiwan) in chip technology that powers AI applications.
  - Autonomous systems: AI is crucial for the development of autonomous systems, including self-driving cars and drones. China has invested substantially in AI for autonomous vehicles and logistics, while the US remains a leader in AI-powered tech companies like Tesla and Waymo.
- *AI-driven global standards and regulations*: Both countries are shaping global AI standards. China pushes for international cooperation aligned with its vision, while the US seeks to protect its technological edge and intellectual property. This competition will determine whose values, ethics, and practices dominate the AI landscape worldwide.
- *AI and data control*: Access to large datasets is crucial for AI development. China benefits from a vast population and lax data privacy laws, giving its firms larger training datasets. The US has stricter regulations but still leverages domestic and international data access. In terms of surveillance and social control, China has deployed AI-powered facial recognition and tracking systems on a massive scale, raising concerns about privacy and potential for social control (which includes its social credit score system).
- *Talent and innovation ecosystems*: Both countries are competing for top AI talent through research institutions, universities, and corporate investments. The US has long led with universities like MIT and Stanford, while China is rapidly expanding its talent pool through public and private investments, making it a rising global AI hub.
- *Geopolitical tensions and AI regulation*: The US-China AI rivalry is driven by trade wars, cybersecurity, and ideological differences. The US accuses China of using AI for surveillance, while China sees US export bans and research limits as efforts to hinder its AI growth. AI can also be used to project soft power and influence other countries, for example, by providing AI-powered solutions for development challenges (Alhammadi et al, 2024). China's emphasis on 'cyber sovereignty' and data governance reflects its ambition to challenge US hegemony in internet governance.
- *Global governance*: The US-China tech war is reshaping global governance in several key ways, influencing international regulations, economic alliances, and digital sovereignty. Both countries are trying to influence global laws, policies, and regulations concerning digital tech through international organizations. The rivalry is playing out in global institutions like the UN, WTO, and ITU (International Telecommunication Union), where both nations are vying for influence over technology regulations and AI governance. China is pushing for a more state-controlled internet model, while the US advocates for a free and open digital space (Khan et al., 2022). China is actively trying to build a 'post-American' global governance regime through initiatives like the BRI and Digital Silk Road. Ironically, however, the digital tech revolution may actually help democratising global governance.

Being the strongest role players in the East and the West, China and the US recognize the centrality of the digital tech sector, especially AI, to their national and global power. This competition for technological supremacy extends beyond economic growth, influencing national security, global influence, and the future of technology governance. AI is acting as a key battleground for technological and geopolitical dominance. While the US has a lead in many areas of AI research and innovation, China is making significant strides through government-backed initiatives and a more flexible regulatory environment, making the AI rivalry and broader 'tech war' between the two nations one of the most defining aspects of twenty-first century geopolitics. Both countries are investing heavily in AI to "win the twenty-first century" (Mirrlees, 2024). From a global perspective, three specific issues require urgent attention: (1) the need for international cooperation to establish norms and standards for the ethical development and use of AI; (2) the rise of AI raises questions about global governance and the need for new frameworks to address the challenges and opportunities presented by this transformative technology; and (3) as competition in especially AI intensifies, the world is moving toward a more fragmented digital order, with significant implications for international trade, security, and innovation (Fingar, 2025).

Despite concerns about the potential negative impacts of AI on sustainable development, such as increased inequalities and environmental issues, it is recognized as a tool that can make a highly significant contribution to the achievement of the SDGs. In fact, the digital revolution as a whole has the potential to act as a catalyst for achieving the SDGs, provided its implementation is inclusive, equitable, and aligned with sustainability principles. Specifically in the case of AI, according to Serge Stinckwich, the use of synthetic data can fast-track countries to hit their SDG targets by the 2030 deadline (Stinckwich et al., 2024). Stinckwich, a computer scientist, is head of research at the United Nations University Institute in Macau, specializing in research and training on the use of digital technologies in addressing global issues. He and other like-minded experts believe governments and institutions can use AI tools responsibly to power their SDG-related research and progress.

One specific example of how AI can be used to speed up progress towards the SDGs is the use of large language models (LLMs), which has already caused a rapid escalation in the amount of data being used to train AI systems (Tsanni, 2024). Synthetic data can help correct biased datasets—for instance, by balancing gender representation in datasets skewed toward one gender. It also addresses data scarcity or missing information, which is especially valuable in fields like medical research where privacy concerns limit access to personal health data. This method is expected to become more widespread. Gartner (2023), a research and consulting firm based in Stamford, Connecticut, predicted that over 60% of the data used to train machine learning models will be synthetic by the end of 2024, which was the case. How it accelerate SDG-achievement is by improving data quality, reducing bias, and enabling more inclusive and effective decision-making across various sectors. In particular, SDG 3, 4, 5, 9, 10 and 16 benefit from this because it ensures more accurate, unbiased, and privacy-compliant data. Synthetic data supports AI-driven solutions that address real-world challenges, ultimately driving progress toward multiple SDGs.

Other AI tools and resources that are making a difference in SDG-related research include:

- *For SDG 14 and 15 (Life below water and Life on land)*: Using AI to track illegal fishing and protect marine biodiversity. Using AI to understand how competition for light between plant species affects agroforests in which cocoa trees are grown alongside other trees and crops, which helps to simulate, design and test systems to optimize cocoa production (Tsanni, 2024). Using Google's AI for wildlife conservation to analyse camera trap data to monitor animal populations and biodiversity.
- *For SDG 2 and 8 (Zero hunger and Decent work and economic growth)*: Using satellite imagery and AI to optimize farming, predict crop yields, and manage food supply chains (e.g., by using Google Earth Engine and IBM Watson Decision Platform for Agriculture) (Gupta et al., 2023b). Using AI-powered tools that provide farmers with insights on pests, soil health, and weather conditions to improve food security (e.g., AgriGPT and PlantVillage). In terms of decent work and economic growth, access to financial services enables entrepreneurship, business growth, and job creation, particularly for SMEs, which are key drivers of employment. Financial inclusion provides individuals and SMEs with access to banking, credit, and insurance, helping them manage financial risks, invest in opportunities, and escape poverty. AI can drive innovation in financial services by relying on non-traditional data to increase access to credit and reduce information asymmetry for borrowers without a credit history (Mhlanga, 2021).
- *For SDG 6 (Clean water and sanitation)*: Using AI to predict floods and issue early warnings in vulnerable regions. Using an AI-powered water risk mapping tool (e.g., Aqueduct from the World Resources Institute) for monitoring water scarcity and pollution. Artificial Neural Networks (ANNs) can predict floods by running hydrological models.
- *For SDG 7 (Affordable and clean energy)*: In Google data centres, DeepMind's AI for Energy Efficiency is being used to optimize energy use, which has already reduced energy consumption by 30%. Google's Solar Mapper uses AI to assess solar energy potential for buildings worldwide. Also, digital twin technology is being used for high-fidelity models in forecasting renewable energy resources (Shen et al., 2021).
- *For SDG 11 (Sustainable cities and communities)*: AI supports the development of smart cities by collecting data and applying rational decision-making systems to achieve sustainable social, environmental, and economic development (Dionisio et al., 2024). Also using machine learning to track and forecast air pollution levels (e.g., BreezoMeter). There are also 'resilient cities network AI tools' that can help cities prepare for climate disasters and infrastructure challenges.

A key question remains: How can the equity problem be addressed through AI? This requires careful consideration of ethical implications, bias and access. Much of the recent progress in developing LLMs has been driven by private firms rather than academic and research institutions. To address this imbalance, public–private partnerships and initiatives can help democratize access to computing infrastructure. For instance, the Swiss International Computation and AI Network, led by the Swiss Federal Institute of Technology in Zurich, provides researchers from low-income regions with supercomputing resources to develop AI tools for global benefit (Tsanni, 2024). Partnering with organizations like Data Science Africa, a Nairobi-based non-profit organization, they empower young Africans to use data science for local problem-solving and to bridge gaps in data and software infrastructure. Closing the digital divide requires AI development platforms to be accessible and open-source, enabling shared resources, datasets and models (Dixon, 2023). This approach fosters inclusivity, reduces resource consumption, and minimizes AI's environmental footprint. Through effective governance, transparency, bias mitigation, and human oversight, AI can be a powerful tool for promoting equity and achieving the SDGs.

Lastly, it is vital that clear directions are given for development cooperation in the area of digital transformation. For instance, digital technology on its own is not inherently inclusive; we need to design it that way. The starting point is to create a foundation environment for the realization of an ethical digital ecosystem (Yoo & Song, 2021). The techno-ethical framework proposed in this chapter could assist with this meaningfully. Additionally, regulations and systems must be in place to ensure the proper functioning of the digital environment, while national capabilities should be enhanced by improving digital literacy through education and developing skilled human resources. Collaboration with various stakeholders is also crucial for addressing global challenges and cross-border issues. To strengthen the digital value chain, emphasis should be placed on fostering entrepreneurship, innovation, and partnerships with the private sector. Furthermore, for the digital transformation to be effective, ICT should be integrated across all areas of development cooperation (such as public administration, transportation, education, health, sanitation and energy), with planning and implementation of donor activities aligned with the 17 SDGs.

In this regard, Zakhozhyi and Ma (2023) identified three particular ways in which digital transformation can drive more sustainable and inclusive development. It is through enabling:

- *Economic development*: Digital transformation enhances information access, stimulates productivity, enables data-driven decision-making, and expands market reach—particularly benefiting SMEs, which have been key to post-COVID-19 economic recovery. Digital platforms and infrastructure should drive economic activity, generate employment, and contribute to poverty reduction efforts.
- *Resilience building*: Digitalization is crucial for both development and crisis management. During the COVID-19 pandemic, those without digital skills or

access to technology struggled to receive proper healthcare. Countries with prior investments in digital infrastructure were more resilient in the crisis. In fragile or crisis-hit areas, digital technologies ensure continuity in services, support effective aid programs, and enhance preparedness.
- *Climate adaptation*: Digital technologies offer innovative solutions to tackle environmental challenges and boost climate resilience, especially for those most vulnerable to natural disasters. With increased computational power and advanced modelling, disaster simulations and recovery plans are now more effective. The vast amounts of available data provide insights into our planet's resources and help build more resilient livelihoods, particularly for at-risk populations.

Specifically, given the strong connection between digital technology and the SDGs, it would be beneficial to enhance collaboration in sectors like Education, Health, and Gender, where digital solutions can have a greater impact. Given the link between digital penetration and economic growth, a certain level of infrastructure is crucial. It is especially important to create an open and secure broadband network environment. The strategy should focus on encouraging private investment alongside Official Development Assistance (ODA) support for partner countries. Finally, it is important to explore how collaboration in digital technology can enhance the digital value chain moving forward. Given that strengthening the value chain is crucial to building a digital ecosystem, it is essential to consider the components of ICT services during the cooperation process carefully. Ultimately, digital transformation forms part of reshaping systemic change in the economy.

References

Alhammadi, A., Alsyouf, I., Semeraro, C., & Obaideen, K. (2024). The role of industry 4.0 in advancing sustainability development: A focus review in the United Arab Emirates. *Cleaner Engineering and Technology, 18*, 1–20. https://doi.org/10.1016/j.clet.2023.100708

Bagwandeen, M. (2021). Don't blame China for the rise of digital authoritarianism in Africa. *LSE*. https://blogs.lse.ac.uk/africaatlse/2021/09/09/dont-blame-china-for-rise-of-digital-authoritarianism-africa-surveillance-capitalism/

Bai, C., Zhou, H. and Sarkis, J. (2023). Evaluating Industry 4.0 technology and sustainable development goals—A social perspective. *International Journal of Production Research, 61*(23), 8094–8114

Bambrough, B. (2024). *Trump confirms bitcoin reserve plans—$15 trillion price boom predicted*. https://www.forbes.com/sites/digital-assets/2024/12/14/trump-confirms-bitcoin-reserve-plans-15-trillion-price-boom-predicted/

Bremmer, I. (2021). The technopolar moment: How digital powers will reshape the global order. *Foreign Affairs, 100*(6), 112–128.

Cheng, J., & Zeng, J. (2024). "Digital Silk Road" as a slogan instead of a grand strategy. *Journal of Contemporary China, 33*(149), 823–838. https://doi.org/10.1080/10670564.2023.2222269

DiNardo, G. (2023). *US falling behind China in tech advancements, Govini report finds*. https://www.defensenews.com/industry/2023/07/17/us-falling-behind-china-in-tech-advancements-govini-report-finds/?utm_source=chatgpt.com

Dionisio, M., De Souza Junior, S. J., Paula, F., & Pellanda, P. C. (2024). The role of digital social innovations to address SDGs: A systematic review. *Environment, Development and Sustainability, 26*, 5709–5734.

Dixon, R. B. L. (2023). A principled governance for emerging AI regimes: Lessons from China, the European Union, and the United States. *AI and Ethics, 3*, 793–810. https://doi.org/10.1007/s43681-022-00205-0

Dolan, M. (2025). *'Golden age' here already, Trump just needs to protect it*. https://www.reuters.com/markets/golden-age-here-already-trump-just-needs-protect-it-mike-dolan-2025-01-22/

Dong-Yeon, K. (2025). How human-centred technology combines innovation and inclusion in Gyeonggi Province, South Korea. *WEF*. https://www.weforum.org/stories/2025/01/human-centred-technology-innovation-inclusion/?utm_source=chatgpt.com

Duthoit, A. (2024). *Tech wars: US-China rivalry for electronics out to 2035*. https://www.coface.com/news-economy-and-insights/tech-wars-us-china-rivalry-for-electronics-out-to-2035#:~:text=American%20companies%20face%20a%20dilemma,-For%20US%20companies&text=At%20the%20same%20time%2C%20despite,in%20the%20strategic%20semiconductor%20segment

Economy, E. (2024). China's alternative order and what America should learn from it. *Foreign Affairs, 103*(3), 8–24.

Erie, M. S., & Streinz, T. (2021). The Beijing effect: China's digital Silk Road as transnational data governance. *Journal of International Law and Politics, 54*(1), 1–92.

European Commission. (2024). *A Europe fit for the digital age: Empowering people with a new generation of technologies*. https://commission.europa.eu/strategy-and-policy/priorities-2019-2024/europe-fit-digital-age_en

Fingar, C. T. (2025). *Global Trends 2025: A Transformed World*. National Intelligence Council.

FOORD. (2024). *Tech titans in focus: Contrasting the US and Chinese tech markets*. https://foord.co.za/insights/the-tech-titans-in-focus-contrasting-the-us-and-chinese-tech-markets

Galvin, S. (2025). *US Navy bans members from using China's DeepSeek AI app out of security fears*. https://www.newsbreak.com/new-york-post-509648/3781641113499-us-navy-bans-members-from-using-china-s-deepseek-ai-app-out-of-security-fears

Gartner. (2023). *Gartner identifies top trends shaping the future of data science and machine learning*. https://www.gartner.com/en/newsroom/press-releases/2023-08-01-gartner-identifies-top-trends-shaping-future-of-data-science-and-machine-learning

Gedeon, J. (2024). Trump unveils $500bn Stargate AI project between OpenAI, Oracle and SoftBank. *The Guardian*. https://www.theguardian.com/us-news/2025/jan/21/trump-ai-joint-venture-openai-oracle-softbank

Geller, E. (2025). *Biden's cyber ambassador urges Trump not to cede ground to Russia and China in global tech fight*. https://www.wired.com/story/nathaniel-fick-us-cyber-ambassador-exit-interview/?utm_source=chatgpt.com

Genspark. (2024). *Top 20 Chinese tech giants by market capitalization*. https://www.genspark.ai/spark/top-20-chinese-tech-giants-by-market-capitalization/d4b60ed3-6e2f-4e7b-95c0-4cfe0d95b577?utm_source=chatgpt.com

Gerken, T. (2025). Be careful with DeepSeek, Australia says—So is it safe to use? *BBC*. https://www.bbc.com/news/articles/cx2k7r5nrvpo

Gertz, G., & Evers, M. M. (2020). Geoeconomic competition: Will state capitalism win? *The Washington Quarterly, 43*(2), 117–136. https://doi.org/10.1080/0163660X.2020.1770962

Glover, E. (2025a). *23 IT Companies in India to know*. https://builtin.com/articles/top-it-companies-in-india

Glover, E. (2025b). *The Stargate Project: Inside the American AI industry's $500B infrastructure bet*. https://builtin.com/articles/stargate-project

Govini Report. (2023). The 2023 National security scorecard: Critical technologies edition. Govini: Arlington, Virginia

Gupta, S., Wang, Y., & Czinkota, M. (2023a). Reshoring: A road to Industry 4.0 transformation. *British Journal of Management, 34*, 1081–1099. https://doi.org/10.1111/1467-8551.12731

Gupta, S., Campos Zeballos, J., del Río Castro, G., Tomi˘ci´c, A., Andrés Morales, S., Mahfouz, M., Osemwegie, I., Phemia Comlan Sessi, V., Schmitz, M., Mahmoud, N., et al. (2023b). Operationalizing digitainability: Encouraging mindfulness to harness the power of digitalization for sustainable development. *Sustainability, 15*, 1–37.

Haeck, P. (2023). EU legislators strike deal on €43B chips plan. *Politico.* https://www.politico.eu/article/eu-legislator-strike-deal-e43-billion-plan-boost-chips-production/

Hemmings, J. (2020). Reconstructing order: The geopolitical risks in China's Digital Silk Road. *Asia Policy, 15*(1), 5–21.

Kasinathan, P., Pugazhendhi, R., Elavarasan, R. M., Ramachandaramurthy, V. K., Ramanathan, V., Subramanian, S., Kumar, S., Nandhagopal, K., Raghavan, R. R. V., Rangasamy, S., & Devendiran, R. (2022). Realization of sustainable development goals with disruptive technologies by integrating Industry 5.0, Society 5.0, smart cities and villages. *Sustainability, 14*(22), 1–31

Khan, S. A. R., Ibrahim, R. L., Al-Amin, A. Q., & Yu, Z. (2022). An ideology of sustainability under technological revolution: Striving towards sustainable development. *Sustainability, 14*, 1–21.

Kochkokin, B. (2017). Chinese big five tech companies gain on U.S. counterparts. *Bloomberg.* https://www.bloomberg.com/news/articles/2017-06-23/chinese-big-five-tech-companies-gain-on-u-s-counterparts

Laidley, C. (2025). *What to expect from the magnificent seven in 2025.* https://www.investopedia.com/what-to-expect-from-the-magnificent-seven-in-2025-nvidia-apple-microsoft-tesla-google-amazon-meta-8765203#:~:text=The%20Magnificent%20Seven%20group%20of,than%2075%25%20the%20year%20before

Lewis, J.A. (2023). *Rethinking technology transfer policy toward China.* https://www.csis.org/analysis/rethinking-technology-transfer-policy-toward-china?utm_source=chatgpt.com

Lundvall, B., & Rikap, C. (2022). China's catching-up in artificial intelligence seen as a co-evolution of corporate and national innovation systems. *Research Policy, 51*, 1–13.

Mearsheimer, J. J. (2021). The inevitable rivalry: America, China, and the tragedy of great-power politics. *Foreign Affairs, 100*(6), 48–59.

Mhlanga, D. (2021). Artificial intelligence in the Industry 4.0, and its impact on poverty, innovation, infrastructure development, and the sustainable development goals: Lessons from emerging economies? *Sustainability, 13*, 1–16.

Mhlanga, D. (2022). Human-centered artificial intelligence: The superlative approach to achieve sustainable development goals in the Fourth Industrial Revolution. *Sustainability, 14*, 1–22.

Mirrlees, T. (2024). The US and China's digital tech war—A new rivalry within and beyond the US empire. *Socialist Register*, 1–36.

Mondejar, M. E., Avtar, R., Diaz, H. L. B., Dubey, R. K., Esteban, J., Gómez-Morales, A., Hallam, B., Mbungu, N. T., Okolo, C. C., Prasad, K. A., & She, Q. (2021). Digitalization to achieve sustainable development goals: Steps towards a smart green planet. *Science of the Total Environment, 794*, 1–20.

Mpofu, F. Y. (2023). Fintech, the Fourth Industrial Revolution technologies, digital financial services and the advancement of the SDGs in developing countries. *International Journal of Social Science Research and Review, 6*(1), 533–553.

Nazzaro, M. (2025). *OpenAI's Sam Altman calls DeepSeek model 'impressive'*. https://thehill.com/policy/technology/5110258-openai-rival-deepseek-ai-model/

Ng, K. (2025). Xi and Putin hold video call after Trump's inauguration. *BBC.* https://www.bbc.com/news/articles/cg7z1rj2vdeo

Ng, K., Drenon, B., Gerken, T., & Cieslak, M. (2025). DeepSeek: The Chinese AI app that has the world talking. *BBC.* https://www.bbc.com/news/articles/c5yv5976z9po

Perrigo, B. (2025). Why Biden is rushing to restrict AI chip exports. *Time.* https://time.com/7206500/biden-ai-chip-export-restrictions/?utm_source=chatgpt.com

Picchi, A. (2025). *What is Stargate, Trump's ambitious AI infrastructure venture?* https://www.cbsnews.com/news/trump-stargate-ai-openai-softbank-oracle-musk/

Pigola, A., da Costa, P. R., Carvalho, L. C., Silva, L. F. D., Kniess, C. T., & Maccari, E. A. (2021). Artificial intelligence-driven digital technologies to the implementation of the sustainable development goals: A perspective from Brazil and Portugal. *Sustainability, 13*, 1–28. https://doi.org/10.3390/su132413669

Polyakova, A., & Boyer, S. P. (2018). The future of political warfare: Russia, the West, and the coming age of global digital competition. *Foreign Policy at Brookings*, 1–19.

Reuters. (2024). *Russia's Putin questions the need for dollar forex reserves, touts bitcoin.* https://www.reuters.com/markets/currencies/russias-putin-questions-need-dollar-forex-reserves-touts-bitcoin-2024-12-04/

Richnák, P., & Fidlerová, H. (2022). Impact and potential of sustainable development goals in dimension of the technological revolution Industry 4.0 within the analysis of industrial enterprises. *Energies, 15*, 1–20. https://doi.org/10.3390/en15103697

Roose, K. (2025). Why DeepSeek could change what Silicon Valley believes about A.I. *The New York Times*. https://www.nytimes.com/2025/01/28/technology/china-deepseek-ai-silicon-valley.html

Schwab, K. (2017). *The fourth industrial revolution.* Penguin.

Sharp, A. (2025). *China condemns U.S. 'coercion' over Panama Canal.* https://foreignpolicy.com/2025/02/07/china-panama-canal-us-trump-bri-mulino/

Shahzad, U., Ferraz, D., Nguyen, H., & Cui, L. (2022). Investigating the spill overs and connectedness between financial globalization, high-tech industries and environmental footprints: Fresh evidence in context of China. *Technological Forecasting & Social Change, 174*, 1–14.

Shen, H. (2018). Building a Digital Silk Road? Situating the Internet in China's Belt and Road Initiative. *International Journal of Communication, 12*, 2683–2701.

Shen, H., Ali, S. A., Alharthi, M., Shah, A. S., Basit Khan, A., Abbas, Q., & ur Rahman, S. (2021). Carbon-free energy and sustainable environment: The role of human capital and technological revolutions in attaining SDGs. *Sustainability, 13*, 1–17.

Sherry, B. (2025). *Trump announces Stargate, a massive new investment in AI.* https://www.inc.com/ben-sherry/project-stargate-massive-new-trump-era-investment-in-ai/91109890

Sparviero, S., & Ragnedda, M. (2021). Towards digital sustainability: The long journey to the sustainable development goals 2030. *Digital Policy, Regulation and Governance, 23*(3), 216–228.

Starrs, S. K., & Germann, J. (2021). Responding to the China challenge in techno-nationalism: Divergence between Germany and the United States. *Development and Change, 52*(5), 1122–1146. https://doi.org/10.1111/dech.12683

Statista. (2025). *Google, Amazon, Meta, Apple, and Microsoft (GAMAM)—statistics & facts.* https://www.statista.com/topics/4213/google-apple-facebook-amazon-and-microsoft-gafam/#topicOverview

Steiner, A. (2023). First regular session of the executive board 2023. *UNDP*. https://www.undp.org/speeches/first-regular-session-executive-board-2023

Stinckwich, S., De Wilde, P., Arora, P., Buarque, F., Chin, Y., Thinyane, M., Fournier-Tombs, E., & Marwala, T. (2024). *Recommendations on the use of synthetic data to train AI models.* United Nations University.

Tirole, J. (2023). Competition and the industrial challenge for the digital age. *Annual Review of Economics, 15*, 573–605. https://doi.org/10.1146/annurev-economics-090622-024222

Tlemsani, I., Zaman, A., Hashim, M. H. M., & Matthews, R. (2023). Digitalization and sustainable development goals in emerging Islamic economies. *Journal of Islamic Accounting and Business Research*, 1–25. https://doi.org/10.1108/JIABR-03-2023-0092

Tsanni, A. (2024). Will Artificial Intelligence help or hinder progress on the SDGs? *Nature.* https://www.nature.com/articles/d41586-024-03959-1#:~:text=Synthetic%20data%20are%20one%20of,environmental%20and%20ethical%20concerns%20remain

United Nations. (2015). MDG Gap Task Force Report 2015. https://www.un.org/millenniumgoals/news.shtml

United Nations Development Programme (UNDP). (2022). *Inclusive by design: accelerating digital transformation for the global goals.* United Nations, New York.

United Nations Development Programme (UNDP). (2025). Background on the goals. *Denmark SDG Accelerator*. https://www.undp.org/sdg-accelerator/background-goals#:~:text=The%20Sustainable%20Development%20Goals%20(SDGs,economic%20challenges%20facing%20our%20world

Vyas-Doorgapersad, S. (2022). The use of digitalization (ICTs) in achieving sustainable development goals. *Global Journal of Emerging Market Economies, 14*(2), 265–278.

Wendling, J. (2024). Charted: Magnificent 7 market cap as a share of the S&P 500. https://www.visualcapitalist.com/sp/charted-magnificent-7-market-cap-as-a-share-of-the-sp-500/

Xiong, Y., & Dai, L. (2023). Does green finance investment impact on sustainable development: Role of technological innovation and renewable energy. *Renewable Energy, 214*, 342–349.

Yoo, S., & Song, Y. (2021). Role of digital technology in achieving the sustainable development goals: Focus on the efforts of the international community. *Journal of International Development Cooperation, 16*(2), 31–57.

Zakhozhyi, V. (2023). *Why you should take a whole-of-society approach to national digital transformation*. https://apolitical.co/solution-articles/en/why-you-should-take-a-whole-of-society-approach-to-national-digital-transformation

Zakhozhyi, V., & Ma, Y. J. (2023). *Three ways digital transformation accelerates sustainable and inclusive development*. https://www.undp.org/blog/three-ways-digital-transformation-accelerates-sustainable-and-inclusive-development#:~:text=Digital%20technology%20brings%20tremendous%20opportunities,also%20widen%20socioeconomic%20gaps%20for

West Versus East: New Capitalism or New Socialism?

4

Abstract

This chapter examines the evolving landscape of capitalism and socialism in the twenty-first century, particularly in the context of West versus East dynamics, and asks whether these different systems are, in fact, merging? The purpose is to analyze the transformations occurring within Western capitalism, including the rise of stakeholder capitalism and surveillance capitalism, and to compare these with the state-led approach in China, questioning traditional boundaries between capitalism and socialism. The chapter discovers that Western capitalism is being remade through a combination of inclusion and utility maximization, with stakeholder capitalism addressing some limitations of shareholder primacy, while surveillance capitalism introduces new concerns about exploitation. On the phenomenon of surveillance capitalism, the chapter highlights its impact on individual freedoms and data privacy, as well as the implications for corporate responsibility and societal welfare. It also examines how China's state capitalism blurs the lines between capitalism and socialism, combining market-driven policies with strong state intervention. Ultimately, the chapter finds that while state capitalism is not the new socialism, the progressive versions of both systems are converging, especially through trends that emphasize economic sustainability and social equity. This progressive synergy has implications for achieving the SDGs, as it can lead to more balanced and inclusive approaches to economic, social, and environmental challenges. However, as identified in this chapter, the resurgence of Trumpian capitalism may disrupt this trajectory.

Keywords

Shareholder versus stakeholder capitalism • New socialism • Trumpian capitalism • SDGs

A. J. van Niekerk, *West to East: A New Global Economy in the Making?*,
https://doi.org/10.1007/978-3-031-93267-0_4

1 Introduction

In an era defined by rapid technological advancements, environmental crises, and growing socio-economic disparities, traditional notions of capitalism and socialism are being challenged. The global economic landscape is a tapestry woven with diverse threads of capitalism and socialism, each evolving and adapting in response to the currents of the twenty-first century. In the West, the conventional model of shareholder capitalism, with its singular focus on profit maximization, is being challenged and reshaped by emerging paradigms such as stakeholder capitalism and surveillance capitalism. These new forms represent a profound shift in corporate priorities, reflecting an increasing awareness of social, environmental, and ethical responsibilities. Juxtaposed against this backdrop is the resurgence of state intervention in the East, exemplified by China's unique model of state capitalism. This hybrid system, blending elements of market dynamism with centralized control, has not only fuelled China's economic ascent but also sparked debates about the future trajectory of global capitalism (Samans & Nelson, 2022). The lines between capitalism and socialism, once sharply defined, are now blurring, prompting a re-evaluation of traditional economic ideologies. Headlines touting China's economic success, the rise of tech giants, and the call for greater corporate responsibility signal a significant shift in the global economic order. The world is witnessing a potential convergence or collision of economic systems, prompting a crucial examination of the evolving dynamics between West and East.

This chapter navigates the complex and evolving terrain of contemporary capitalism and socialism, scrutinizing the transformations occurring within these systems and their implications for global development. It the context of the remaking of Western capitalism, it provides a comprehensive analysis of the transformations underway in both Western and Eastern economic models. These models challenge traditional shareholder-focused approaches and raise critical questions about corporate responsibility and economic exploitation. The phenomenon of surveillance capitalism, with its implications for individual privacy and data ethics, will also be scrutinized, raising crucial questions about the balance between innovation and regulation (Zuboff, 2019). Simultaneously, the chapter explores the rise of state capitalism in China, a hybrid system that blends market-driven policies with strong state intervention. This unique approach blurs the conventional boundaries between capitalism and socialism, prompting a re-evaluation of established economic paradigms. A comparative lens will be employed to also examine the similarities and divergences between state capitalism and socialism, shedding light on the unique characteristics of each system and their distinct approaches to governance, ownership, and market mechanisms. Furthermore, the chapter investigates the convergence and synergy between capitalism and socialism, assessing their combined impact on the United Nations' SDGs.

The overarching aim is to foster a deeper understanding of the contemporary economic landscape, characterized by the dynamic interplay between capitalism and socialism, particularly in the context of West versus East dynamics. By examining these evolving models, the chapter seeks to provide insights into the challenges and opportunities that lie ahead, as nations grapple with issues of economic growth, social equity, and environmental sustainability. This analysis contributes to understanding the contemporary shifts in global economic systems and their broader societal and political ramifications. The discussion will be grounded in a theoretical framework that draws upon insights from political economy, organizational behaviour, and ethics. It involves Marxist critiques of shareholder primacy and exploitation, stakeholder theory, and theories of state capitalism. This interdisciplinary approach will facilitate a holistic analysis of the multifaceted issues at hand, providing a nuanced understanding of the forces shaping our economic future.

At its core, this chapter seeks to answer this primary question: Are capitalism and socialism merging, and what are the implications of this convergence for global economic and social progress in the twenty-first century? Two critical, but divergent issues, are particularly investigated: (1) the potential for progressive versions of capitalism and socialism to promote sustainable and inclusive development; and (2) the impact of the resurgence of Trumpian capitalism on the evolving dynamics between these systems. The chapter holds significant implications for policymakers, business leaders, and scholars alike. It informs debates about the future of corporate governance, the role of the state in the economy, and the pursuit of sustainable development. Specifically, its exploration of China's state capitalism sheds light on an increasingly influential economic model that is reshaping the global balance of power.

2 Remaking Western Capitalism: From Shareholder Capitalism to Stakeholder Capitalism and Surveillance Capitalism

The Western world is experiencing a profound transformation in its capitalist structures, adapting to the evolving global landscape. This shift is largely characterized by a move away from a traditional shareholder-focused model to new inclusive paradigms such as stakeholder capitalism, as well as incorporating elements of surveillance capitalism. Marking a significant departure from established norms, these shifts reflect changing economic, technological, and societal forces that are redefining corporate priorities and power structures.

Shareholder capitalism, dominant since the mid-twentieth century, prioritizes maximizing profits and shareholder value, often to the exclusion of other considerations. A company's success is measured primarily by the size of its profits and how much it maximizes shareholder value (Kim, 2022). As a good portrayal of traditional capitalism (an economic system based on the private ownership of the means of production and their operation for profit), shareholder capitalism—focussing on the dominant corporate form—involves legally independent companies that raise (and pool) capital from multiple shareholders with limited liability, supported by an open stock market that allows for the free trading of shares. It is rooted in the belief that prioritizing individual company financial returns protects shareholders and optimizes resource allocation. It has an excessive short-term focus where managers are encouraged to focus on short-term share price movements, which places pressure on CEOs and managers to meet quarterly earnings targets. This can adversely affect the long-term health and sustainability of the business. Because it is characterized by private property, capital accumulation, competitive markets, commodification, wage labour, and a strong emphasis on innovation and economic growth, such a model of the economy tend to experience business cycle fluctuations of economic growth followed by recessions (Silverman & Ingram, 2017).

Having dominated corporate governance for decades, shareholder capitalism has led investors to exacerbate the focus on individual company returns by following Modern Portfolio Theory, which emphasizes relative performance against peers rather than absolute performance and common sustainability guardrails (Mazzucato, 2024). Mainly short-term gains drive business decisions. Over time, it developed into a situation where shareholders and their interests are dominant globally, driving 'supercapitalism' and resulting in the 2007–09 global financial crisis. Similar to neoliberal capitalism, contemporary shareholder capitalism has evolved into a 'financialized' form, where shareholder value increasingly dictates corporate investment, raising concerns about the efficacy of the corporate structure and equity market in facilitating investment (Samans & Nelson, 2022). In addition, growing concerns about economic inequality, environmental degradation, and corporate accountability have caused a rethink of shareholder capitalism, which has led to the rise of stakeholder capitalism. Arguably, the most significant criticism of shareholder capitalism is that it often leads to companies being accountable to an abstract ideal of shareholder value maximization rather than to real people with a stake in the company's long-term success.

Stakeholder capitalism asserts that companies should have a more holistic/comprehensive approach and serve not just shareholders but also employees, customers, communities, and the environment. It is an economic model that prioritizes the interests of all stakeholders—not just shareholders (Schwab & Vanham, 2021a). This means companies consider the well-being of employees, customers, communities, and the environment alongside financial profits. This model promotes sustainable and ethical business practices rather than focusing solely on short-term profits for investors. Stakeholder capitalism brings a shift in focus by broadening the corporate purpose beyond maximizing shareholder value to include the interests of all stakeholders. Technological, environmental, geopolitical, and socio-economic changes are compelling corporate boards and management teams to move towards this form of capitalism. More specifically, this means:

- *Technological transformation*: The scaling effects of new digital technologies and business models are contributing to the rise of industrial policies.
- *Environmental concerns*: There has been a major shift in social attitudes, energy markets, regulatory agendas, and consumer and investor preferences regarding the need to take urgent action on climate change (Samans & Nelson, 2022). Firms are required to integrate climate change considerations into corporate

governance, strategy, risk management, and disclosure. Environment, Social and Governance (ESG[1]) scores is an example of this, which assesses a company's performance across these three key areas.

- *Geopolitical factors*: The growing multipolarity of international relations and the return of overt great power rivalry are contributing to the ongoing plurilateralization of the world economy. Uncertainty and complexity are rising, requiring MNCs to take a more deliberate approach to assessing risks.
- *Socio-economic factors:* Societal expectations of corporations are shifting as widespread concern about automation, trade, climate change, inequality, immigration, and corporate ownership of personal data rises (Freeman et al., 2007). There is a deficit of trust in corporations in many countries, as well as growing debate about whether they contribute sufficiently to the broad-based gains in living standards that come from inclusive economic growth.
- *Growing materiality of ESG&D issues*: The trends mentioned above are increasing the materiality of ESG and Data stewardship (ESG&D) issues for corporations in almost every industry sector. ESG&D risks and opportunities are having a growing impact on the financial condition or operating performance of companies.
- *The COVID-19 pandemic*: The COVID-19 pandemic has highlighted the challenges of inequality and vulnerability within company value chains and the communities where they operate. Thus, the post-pandemic world brought about a higher level of conscientiousness.

Given heightened sensitization, younger generations, particularly, are advocating for sustainability and inclusion on corporate agendas. They are more and more eager to have their voice heard, as workers, consumers, and investors, and are sending the message that sustainability and inclusion should be much higher on corporate priority lists (Zanoni & Pitts, 2022). This is related to the fact that Gen Z is characterized by a transition from the concept of ownership to the concept of use, which leads to the spread of the sharing economy (e.g., Uber and Airbnb) and stakeholder capitalism. They value access relatively more than ownership, hence the need to preserve and develop existing economic, social and environmental assets so that sustainability becomes the primary goal of innovation (Van Niekerk, 2022). Gen Zs believe that wealth should perform a social function.

Stakeholder capitalism is aligned with the values and expectations of younger generations, who prioritize social and environmental responsibility. They have a strong desire for systemic change, whereby stakeholder capitalism is seen as a means of solving modern environmental and social problems in close connection with the goals of sustainable development. Young investors and consumers demonstrate increased environmental concerns, which is driving higher returns of ESG

[1] ESG takes the holistic view that sustainability extends beyond just environmental issues. An ESG score is a way to assign a quantitative metric, such as a numerical score or letter rating, to ESG efforts. ESG investing refers to how companies score on these responsibility metrics and standards for potential investments.

assets. In a study by Allen et al. (2009), it was found that younger managers in line for upper management positions are increasingly more concerned with maintaining continuity of the firm and may influence a firm's investments in a way that increases shareholder value. This is in the context of rejecting the separation hypothesis, which distinguishes between shareholders and other stakeholders.

Given that stakeholder capitalism recognizes and rewards contributions to the value creation process by all stakeholders, it is often called 'inclusive capitalism'. Both models challenge traditional shareholder capitalism by advocating for a more equitable distribution of economic benefits and a focus on long-term sustainability rather than short-term profits (Michaels, 2020). Both aim to create a more just and sustainable economic system where the benefits of economic growth are shared more broadly. While they may have slightly different emphases, they are closely related and often used interchangeably. Whereas stakeholder capitalism is a corporate-level shift, ensuring that businesses serve all stakeholders rather than just shareholders, inclusive capitalism is a broader economic movement aimed at reshaping capitalism to reduce inequality and ensure that prosperity is widely shared. Stakeholder capitalism is therefore a key pillar of inclusive capitalism, as businesses must adopt stakeholder principles for a truly inclusive economy to emerge. Inclusive capitalism is a moral form of capitalism that actively works to reduce disparities in wealth and opportunity. Both stakeholder and inclusive capitalism are trying to solve the same problem, which is the negative effects caused by solely focusing on shareholder profit. In line with stakeholder capitalism, "inclusive capitalism is fundamentally about creating long-term value that benefits all stakeholders—businesses, investors, employees, customers, governments, communities, members of society, and the planet" (Coalition for Inclusive Capitalism, 2025: 1). Accordingly, a capitalism that is inclusive:

- generates fair and sustainable growth that empowers individuals and meets societal needs;
- enables inclusive participation in the economic system, ensuring fair compensation for contributions, and correcting economic inequalities by improving access to capital, education, and opportunities;
- promotes social mobility, ensuring that all individuals can pursue and achieve opportunities regardless of socio-economic background, gender, ethnicity, religion, or age; and
- ensures that short-term gains do not come at the expense of future generations or allow one party to benefit unfairly at the cost of others.

Surveillance capitalism is another contributor to remaking capitalism, especially in the West, since the 2000s. It came about as advertising companies, led by Google's AdWords, saw the possibilities of using personal data to target consumers more precisely. Coined by Professor Shoshana Zuboff (2019: 8), surveillance capitalism "claims human experience as free raw material for translation into behavioural data". Companies extract, analyse, and monetize personal data to influence behaviour and drive profits. The data is captured through surveillance and then analysed—often without explicit consent—to predict and modify consumer behaviour for profit. The rise of digital technologies and platforms has accelerated surveillance capitalism, raising concerns about privacy, ethical data usage, and the potential for manipulation.

Using computational tools to predict behaviour, this form of capitalism is seen as representing a new economic order/system that claims human experience as free raw material for hidden commercial practices of extraction, prediction, and sales (Alshamy et al., 2024). It contains, what some call, a 'parasitic economic logic' in which the production of goods and services is subordinated to behavioural modification. While increased data collection offers benefits like self-optimization, societal improvements (e.g., smart cities), and enhanced services (e.g., web applications), its expansion under capitalism raises concerns. As more aspects of social life become subject to data collection and processing, issues related to privacy, societal control, and vulnerability become more significant. Tech giants like Google, Meta and Amazon dominate this space, leveraging vast amounts of user data for targeted advertising, predictive analytics, and AI-driven personalization. Perhaps the only gain to everyone is that it has fuelled rapid innovation and convenience.

Capitalist economic pressures are accelerating online connectivity and surveillance, allowing corporate entities to increasingly infiltrate social spaces for profit generation and behavioural regulation (Ullah et al., 2021). In the twenty-first century, the value of personal data surged with the rise of targeted advertising, making data a highly prized commodity (becoming 'digital gold'). As a result, the growing cost of acquiring personal data has restricted access primarily to society's wealthiest individuals (especially the owners of big tech companies). In the West, this phenomenon is distinct from government surveillance, although the two can be mutually reinforcing (as in China). Surveillance capitalism is one area where there could be an overlap between the economic systems being developed in the West and in the East, especially in China. Despite the political and economic systems of the West and China being different, there are a number of commonalities—which is of serious concern—when it comes to surveillance capitalism:

- *Behavioural profiling and prediction*: Both systems use rapidly improving AI and big data to analyse and predict consumer behaviour, influencing purchasing decisions, media consumption, and online interactions. Targeted advertising and personalized recommendations are central in both models.
- *Data as a core asset*: Both rely on extensive data collection as a primary driver of economic and political power. In the West, corporations like Google and Meta monetize user data for advertising and behavioural prediction. In China, companies like Alibaba and Tencent also collect vast data, but with stronger government oversight, directives and integration into state objectives.
- *Integration with governance*: Western surveillance capitalism influences governance indirectly, as tech giants shape policy debates, lobby governments, and

control information flows. In China, surveillance capitalism is closely intertwined with state-led governance, contributing to social credit systems and public security initiatives.
- *Algorithmic influence on public opinion*: Both use algorithm-driven content delivery systems that can amplify certain narratives and shape public discourse. Social media platforms in both regions have been criticized for fostering echo chambers, misinformation, and political polarization.

The increasing convergence of data-driven economies suggest that both models share fundamental mechanisms of surveillance, control, and economic exploitation of personal data. The key difference is that, in the West, surveillance capitalism is primarily market-driven, with corporate control over data, while in China, it is state-driven, with the government leveraging corporate data for broader social and political control (Lin & Milhaupt, 2023). The effect seems to be strikingly similar: "surveillance capitalism is an assault on human autonomy" (Zuboff, 2019: 8). What started as a tool for advertising has become a threat to freedom and democracy (Kavenna, 2019).

Zuboff (2019) contrasts industrial capitalism, which relied on mass production and was interdependent with its consumers and workforce, with surveillance capitalism, which exploits dependent populations that neither serve as its employees nor consumers and remain largely unaware of its operations (i.e., invisible collaborative arrangements with state security apparatuses). Showing the transition and change, Table 1 provides a summarized comparison of capitalism's evolution in the West. Scholz (2016) has raised concern that companies recruit people as informants for surveillance capitalism. Mosco (2014) also warned about the consequences of surveillance capitalism being the practice of selling customer and subscriber data to advertisers. Notably, he observed that the surveillance state operates alongside it, with both systems increasingly merging.

What Table 1 shows is the diverse impact of innovation in how capitalism evolved in the West. Some of the limitations of shareholder capitalism has been addressed by stakeholder capitalism, such as being overly profit-driven and balancing it with a more holistic focus on adding value to all stakeholders (those affected by a company), and not just shareholders. However, the development of surveillance capitalism shows, again, the ugly face of capitalism: the exploitation by the rich/powerful of the poor/disempowered users. As in the case of the increasing gap between the rich and the poor, it is the gain of some (minority) at the expense of others (majority). This reinforces concerns about the sustainability of neoliberal 'supercapitalism' and why we need inclusive capitalism. If there is not sufficient moral responsibility, the drive for riches and benefits distorts the good intensions of capitalism (free markets). It becomes a monster (predatory capitalism); the manipulation of the masses by the powerful, which might in effect not be that far different from the socialism of China (currently expressed as state capitalism). The balancing factor of stakeholder capitalism thus becomes crucial. Clearly, there are many forms of capitalism with different impacts on the environment and society. At the heart of the issue, the West has a choice to make: to either allow

Table 1 Components of the evolution of capitalism in the twenty-first century

Aspect	Shareholder capitalism	Stakeholder capitalism	Surveillance capitalism
Core principle	Maximizing shareholder value and profits	Balancing the interests of all stakeholders, not just shareholders	Monetizing personal data as a primary economic activity
Primary objective	Profit maximization and shareholder returns	Sustainable value creation for shareholders, employees, customers, communities, and the environment	Data extraction, behavioural prediction, and profit from targeted advertising
Main beneficiaries	Shareholders (investors)	Shareholders, employees, customers, suppliers, and society at large	Tech corporations, advertisers, and AI-driven decision-makers
Governance model	Board and executives prioritize shareholder interests	Corporate governance includes ESG considerations	Algorithms, AI systems, and big data analytics drive decision-making
Market orientation	Free-market, competition-driven economy	Market economy with ethical, social, and environmental responsibilities	Data-driven markets where consumer behaviour is predicted and influenced
Economic focus	Short-term financial gains, stock price growth	Long-term value creation, corporate responsibility, ESG impact, inclusive growth	Data commodification, predictive analytics, and digital surveillance
Role of consumers	Buyers of goods and services	Active participants in a responsible and ethical economy	Users who unknowingly provide data that is monetized
Employee relations	Focus on efficiency, cost-cutting, higher profits, and shareholder returns	Emphasis on fair wages, working conditions, and employee well-being	Workers as data sources, often subjected to algorithmic management
Government role	Minimal intervention, deregulation preferred	Stronger regulatory and ethical frameworks for corporate responsibility	Calls for privacy laws and digital rights protections
Ethical considerations	Profit-first mentality, potential for exploitation	Ethical corporate behaviour, environmental and social accountability	Privacy concerns, data security, and ethical AI dilemmas
Environmental impact	Often disregarded unless it affects profits	Integrated into corporate strategy through ESG initiatives	Indirect—data centres and AI systems have high energy consumption

(continued)

Table 1 (continued)

Aspect	Shareholder capitalism	Stakeholder capitalism	Surveillance capitalism
Key industries	Finance, manufacturing, consumer goods	Renewable energy, sustainable businesses, ethical investing	Big tech, digital platforms, AI-driven companies
Examples of companies	ExxonMobil, JPMorgan, General Electric	Patagonia, Unilever, Tesla	Google, Facebook (Meta), Amazon, TikTok

Sources Mazzucato (2024), Schwab and Vanham (2021a), Ullah et al., (2021), Samans and Nelson (2022)

profit-making to drive innovation or to redirect innovation more towards economic sustainability—not divorced from the profit-drive, but not subordinate to it.

The remaking of Western capitalism is a mixture of inclusion and utility maximization. The way in which capitalism is evolving is a reflection of significant adaptations of Western capitalism in response to twenty-first century challenges and opportunities. This shift is largely characterized by a move away from traditional shareholder capitalism towards more inclusive models. Stakeholder capitalism reflects a growing awareness of the social and environmental responsibilities of businesses, while surveillance capitalism underscores the increasing importance of data as a valuable asset in the digital age (Alshamy et al., 2024). Put differently, Western capitalism is undergoing a transformation from shareholder to stakeholder capitalism, as well as incorporating elements of surveillance capitalism. In the remaking of Western capitalism, the shift towards stakeholder capitalism took significant shape when ESG factors started to be integrated in the corporate world (in part, due to SDG-influence), together with increased awareness about the rise of surveillance capitalism and its impact on individual freedoms and democracy.

Specifically, calls to remake capitalism suggest a need to revolutionize the purpose of the firm (i.e., repurposing business). Integrated reporting (of both profit and sustainability indicators) has become a valuable tool for starting to embed corporate transparency, accountability and fairness in corporate management (Alexander et al., 2020). Addressing declining trust in capitalism requires a change in business norms and culture. In addition to this, the state's role as a market shaper has become essential because a collaborative form of stakeholder value is needed. In summary, the transformations that are remaking capitalism, include:

- *Shift from short-term profits to long-term sustainability*: Businesses are integrating ESG goals and adopting more responsible, transparent governance.
- *New power centres*: Tech corporations wield unprecedented economic and political influence through data ownership and AI-driven decision-making.
- *Regulatory challenges and ethical debates*: Governments struggle to regulate big tech, enforce privacy laws, and balance corporate innovation with consumer rights.

- *Worker and consumer agency*: As data becomes a key asset, questions of digital labour rights, fair wages, and data ownership are gaining prominence.
- *Globalization*: The interconnected global economy amplified the impact of corporate actions, making stakeholder considerations more urgent.

These changes represent a broader restructuring of economic power. While stakeholder capitalism aims to align business with social good, surveillance capitalism challenges traditional capitalism by placing data and AI at the centre of wealth creation. This transition through innovation will increasingly shape the future of global markets, democracy, and personal freedoms. The remaking of Western capitalism reflects broader societal changes, including technological innovation, environmental crises, and demands for social justice (Boustanifar, 2023). While stakeholder capitalism and surveillance capitalism represent different trajectories, they both highlight the need for a more equitable and sustainable economic system. However, the tension between profit motives and ethical responsibilities remains unresolved. Importantly, beyond these different forms of capitalism, several other key factors have played a critical role in reshaping and evolving Western capitalism in the twenty-first century, such as:

- *Inequality and the social contract*: Rising income disparities, worker precarity, and social unrest have fuelled calls for fairer economic systems and policies, influencing government interventions and corporate governance models. Growing wealth inequality is driving debates about tax reform, universal basic income (UBI), and corporate responsibility. Declining trust in institutions, including corporations and governments, is urging demands for greater accountability, transparency, and to re-engineer social innovation.
- *Climate change and ESG investing*: Growing pressure for businesses to adopt sustainable practices and align with climate goals has led to a shift in corporate priorities and investment strategies (e.g., using ESG-criteria). Global environmental challenges are forcing businesses and governments to rethink resource use, production methods, and energy systems. There is also a growing recognition of the importance of employee well-being, mental health, and work-life balance in corporate strategies.
- *Automation and AI-driven economies*: The increasing role of AI, blockchain, IoT, robotics, and automation is transforming industries, redefining labour markets, and raising ethical concerns about employment and data ownership. In addition, the dominance of platforms like Amazon, Uber, and Airbnb has redefined how goods and services are exchanged, often bypassing traditional regulatory frameworks. Techno-nationalism and protectionism are have become further concerns in platform governance and data localization.
- *Regulatory and policy shifts*: Governments and international institutions are playing a more active role in shaping capitalism through antitrust measures, data privacy laws (e.g., GDPR and CCPA), taxation policies, and labour protections. Policies like carbon taxes, emissions trading systems, and green energy mandates are applying pressure on companies to adopt sustainable practices.

- *Supply chain reconfiguration and digital transformation*: The COVID-19 pandemic and geopolitical tensions (e.g., US-China rivalry) have exposed vulnerabilities in global supply chains, prompting a shift toward regionalization and reshoring. The pandemic accelerated digital transformation, remote work, and e-commerce while exposing vulnerabilities in healthcare and social safety nets. Rapid technological change is creating a skills gap, i.e., a mismatch between workforce skills and job requirements, prompting a focus on reskilling and lifelong learning.
- *Civic and consumer activism*: Consumers and social movements are demanding greater corporate accountability, ethical and transparent business practices, and sustainability, influencing corporate decision-making and market trends. Movements like #MeToo, Black Lives Matter, and climate activism are pressuring corporations to address social justice and environmental issues.
- *Ethical shifts, reimagining capitalism, and global shifts*: There is growing interest in alternative economic models, such as doughnut economics, degrowth, and post-capitalism, which prioritize sustainability and equity over endless growth. Firms are increasingly being asked to define their purpose beyond profit, aligning their operations with broader societal goals. On the other hand, Western companies need to find new ways to compete better against the East, especially as Chinese tech companies are expanding globally to contest the power of the West, incorporating transnational capital and globalization strategies.

The remaking of Western capitalism is being driven by a confluence of technological, social, environmental, and political forces. While shareholder, stakeholder, and surveillance capitalism represent key frameworks, the broader evolution is shaped by factors like technological disruption, globalization, regulatory changes, and shifting societal values (Zanoni & Pitts, 2022). These dynamics are pushing capitalism toward a more complex, interconnected, and contested future, where the balance between profit, purpose, and sustainability remains a central challenge. There is increasing recognition in the West that strong and sustained value creation depends on rigorous understanding and active management of environmental, social, and governance considerations (De Jong, 2021). Hence, a political economy based on stakeholder value requires institutional innovations to ensure more equitable distribution of value and sustainability of its creation. Particularly the focus on 'collective value creation' is bringing the 'purpose paradigm' to the centre, which emphasizes that *purpose* rather than utility or profit maximization is the key driver in stakeholder-oriented capitalism. At the heart of the matter, this requires innovative business models and financing mechanisms essential to developing and scaling new technologies, products, and services that mitigate ESG risks and create more inclusive and sustainable markets.

3 New Forms of Capitalism and New Socialism: Are They Merging from East to West?

China's success has blurred the traditional boundaries between capitalism and socialism in significant ways, making state capitalism appear to be the new form of socialism. While the two systems have distinct origins, goals, and mechanisms, there are areas where they overlap or converge, particularly in how governments intervene in economies to achieve social or strategic objectives. For instance, some believe the opportunities created by surveillance capitalism could be harnessed by socialist planning. However, a challenge for this new kind of socialism would be to find a form of democratic control that does not reproduce new kinds of exploitation and alienation. It is therefore important to first understand the meanings of state capitalism and socialism in the traditional sense and then investigate its possible morphing in the twenty-first century.

Sperber (2019) points out that state capitalism underwent significant historical evolution over time. Early discussions (late nineteenth to twentieth century) on state capitalism emerged from Marxist-inspired writings. Friedrich Engels and Wilhelm Liebknecht argued that state ownership of industries did not equate to socialism but was instead a form of state capitalism, as the state remained under bourgeois control. They focused on the interpenetration of state and capital, particularly in how the state functioned as both an economic actor and regulator. An example of this was the growing economic role of the state in capitalist societies, particularly in industries like railways, telegraphs and postal services. During this period—between the 1890s to the 1920s—Vladimir Lenin (1917) argued that state capitalism was a necessary transition phase towards socialism, particularly in the Soviet Union after the 1917 Revolution. Karl Kautsky (2019) described state capitalism as a form of state-monopoly capitalism (in the context of imperialism), where governments collaborated with large companies to control markets.

From the mid-twentieth century (1930s–1970s), Friedrich Pollock (1936) introduced the idea of a "state-managed capitalism", arguing that economic planning was overtaking free-market mechanisms. It described the 'statification' of economies during wartime and under fascist regimes, where the state took control of production and distribution. Nikolai Bukharin (1925) theorized that advanced capitalist economies were transforming into "state-capitalist trusts", with the state becoming the supreme economic regulator. Pollock, influenced by the Nazi economy, described "totalitarian state capitalism", where the state controlled economic life through political means. The Soviet Union—the world's first socialist state—was seen as a state-capitalist system where the state acted as the collective capitalist, exploiting workers similarly to private capitalists as bureaucratic elites functioned like private capitalists, accumulating surplus for the state.

The post-World War II period saw the rise of the welfare state and Keynesian economics in Western countries. The symbiotic relationship between the state and large private monopolies in Western economies became a major focus area. The state was seen as promoting the interests of private capital through subsidies, regulation, and public enterprises, leading to the concept of "state-monopoly capitalism" by Paul Sweezy (1946). He pointed to how Western governments (especially the US) used state power to support private monopolies. In China, Charles Bettelheim (1985) analysed Maoist economic policies as an alternative form of state capitalism, driven by state-controlled production but distinct from Western models. State capitalism also became a development strategy in Third World countries like Brazil, Algeria, and Zambia, where the state played a central role in economic planning and industrialization. It was seen as a 'third way' between dependency on Western capitalism and socialist revolution, with the state acting as both an accumulator of capital and a catalyst for private accumulation (Howard & King, 2001). With the decline of the Soviet Union and the global shift towards neoliberalism, interest in state capitalism waned, resulting in a decline in Marxist discourse.

The 2007–09 GFC revived interest in state capitalism, led this time not by Marxist scholars but by business consultants, think tanks, and financial journalists. Scholars analysed how governments manage capitalism differently. The discourse shifted towards a free-market perspective; analysing state capitalism in relation to emerging economies. The resurgence of state capitalism in emerging economies like China, Russia, and Brazil are examples of this, where the state plays a significant role in economic activities through state-owned enterprises (SOEs), sovereign wealth funds (SWFs),[2] and industrial policies. Earlier studies examined macro-institutional dynamics (i.e., state capitalism as a systemic phenomenon), whereas contemporary literature tends to focus on micro- and meso-level analyses, which considers the narrower aspects of state involvement in the economy (e.g., governance of SOEs and policies). The distinction between state capital (SOEs) and statism (state influence over private actors) has been proposed as a framework to navigate the concept today (Bremmer, 2009). The early focus was systemic and critical (Marxist critique), emphasizing state capitalism's role in power dynamics and economic accumulation. Diverging from Marxist roots, the modern resurgence emphasize pragmatic, market-based analyses—adopting a free-market perspective.

This 'new state capitalism' is characterized by a mix of state ownership and market mechanisms, with the state acting as a major economic player while allowing private enterprise to flourish. It can be seen as a continuum between liberal market economies and coordinated market economies. It explores how governments like China, Russia, and the Gulf States use state-controlled enterprises for strategic goals. Another focus is the key role that the state plays in driving innovation (e.g., funding tech developments in Silicon Valley and in China and India), countering the idea that only the private sector drives economic progress (Mazzucato, 2024). This historical progression of state capitalism highlights the versatility of the concept and its adaptation to different political, economic, and intellectual contexts over time.

[2] SWFs are state-controlled investment funds that invest in both domestic and international markets (e.g., China Investment Corporation, Norwegian Government Pension Fund Global, and Abu Dhabi Investment Authority).

State capitalism can thus be described as an economic system in which the state plays a dominant role in guiding and controlling the economy, often through: (1) state ownership of capital, whereby the state owns and operates capitalist enterprises (referred to as 'state capital'); and (2) state influence over private actors, whereby the state exerts significant control or influence over private economic actors through mechanisms like regulation, subsidies, taxation, and industrial policy (referred to as 'statism') (Alami & Dixon, 2023). In this system, the government actively engages in business and commercial activities for profit, with the means of production operated through state-owned enterprises. This includes capital accumulation, centralized management, and wage labour. The concept also extends to state control over corporatized government agencies that function with business-management principles, as well as publicly traded companies where the state holds a controlling stake (Sweidan, 2021). Under state capitalism, the state uses and controls the free-market system—often through ownership of key industries, strategic investments, and regulation—to protect its political regime by leading economic activities. This regime dominates the resource allocation mechanism and the resources to guarantee its persistence.

In a state-capitalist country, the government exerts control over the economy, functioning similarly to a large corporation by extracting surplus value from the workforce and reinvesting it into further production. This classification applies regardless of the state's political ideology, even if it is officially socialist. The Soviet Union, its Eastern Bloc allies, and Maoist China operated under state-capitalist systems. Today, the modern economies of China and Singapore are seen as hybrids of state capitalism and private capitalism (Gonzalez-Vicente & Cheng, 2025). The Chinese government controls major industries (e.g., energy and telecommunications), while allowing private businesses to thrive in other sectors. In Singapore, the government invests heavily in SOEs and strategic sectors while fostering a competitive private sector. In Norway, the state manages significant resources (e.g., oil wealth through Statoil) while maintaining a market-based economy. State capitalism is often associated with economic nationalism and strategic control over resources and industries deemed vital to national interests. In essence, it is a private capitalist economy that is regulated and controlled by the state, involving austere economic planning and intervention.

There are a number of different forms of state capitalism; it is not monolithic. Cross-national variations can be identified by positioning countries along different dimensions (e.g., state ownership, statism, and government threat to the private market sphere). Depending on the levels and varieties of state influence, state capitalism exists along a spectrum where the intent and extent of state intervention varies. Apart from classic state capitalism, which involves the state directly owning and managing enterprises across key industries, state capitalism has also found expression in the following diverse ways:

- *Developmental state capitalism*: The state plays an active role in guiding economic development, often through industrial policy, subsidies, and strategic investments. The state plays a central role in economic planning and industrialization, often as a 'third way' (Lin et al., 2021). The state acts as both an accumulator of capital and a catalyst for private accumulation. Examples include Japan (post-World War II), South Korea (1960s–1980s), Singapore, and Brazil under military rule, where the state formed alliances with domestic and foreign capital to promote industrialization.

- *Authoritarian/totalitarian state capitalism*: The state exerts control over major economic sectors while maintaining a centralized political system, often suppressing dissent and limiting private-sector autonomy. Examples include China, Russia, Gulf monarchies (e.g., Saudi Arabia, UAE), and Nazi Germany, where the state controlled the economy through political means, reducing private capitalists to 'rentiers'.
- *State-led mixed economy*: A blend of free-market capitalism and state intervention, where the government actively participates in key industries but allows private enterprise to flourish in others (Musacchio et al., 2015). Examples include Germany (with state-owned banks and industries) and Nordic countries (strong welfare state with state-owned enterprises). A variation is hybrid forms of state capitalism where the state owns either a majority stake (equity position) in enterprises but allows minority private ownership (e.g., Petrobras in Brazil), or holds minority stakes in private firms through investments by development banks, pension funds or SWFs—influencing their operations without full control (e.g., Vale in Brazil and Capitaland in Singapore).
- *Resource-based state capitalism*: The government controls and profits from natural resource extraction, often using revenues to fund public services or maintain political stability (Wright et al., 2021). Examples include Saudi Arabia, Norway (with its oil-based sovereign wealth fund), and Venezuela (PDVSA).
- *Strategic state capitalism*: The state intervenes selectively in industries critical to national security, technological advancement, or geopolitical strategy (Alami & Dixon, 2023). Examples include China's semiconductor and AI investments, US defence industry support, and France's aerospace sector.
- *Neoliberal/monopoly state capitalism*: A paradoxical form where the state facilitates and protects large corporations while adopting a pro-market ideology. Acting as a servant of private capital (ensuring its accumulation and stability), the state forms a symbiotic relationship with large private monopolies, supporting them through subsidies, regulation, and public enterprises. Examples include US government bailouts (e.g., 2007–09 GFC) and privatization under Thatcherism in the UK.
- *New state capitalism in the twenty-first century*: The rise of emerging economies like China, Russia, and Brazil in the late 2000s, following the GFC where the state combines ownership of key industries with market-oriented policies. SOEs, sovereign wealth funds, and national champions play a significant role in the economy. The state influences private economic actors through industrial policies, subsidies, and regulation. Examples include SOEs in China that coexist with a growing private sector, with the state guiding economic development

through industrial policies; SOEs in Brazil like Petrobras operate alongside private firms, with the state playing a significant role in strategic sectors; and the state in Singapore using SWFs (channelling public wealth into stock markets) and SOEs to drive economic growth while maintaining a market-oriented economy.

Socialism is a political and economic system wherein property and resources (the means of production) are owned in common, typically controlled by the government (National Geographic, 2025). It is based on the idea that common or public ownership of resources leads to a more equal society. In the socialist view, notions of individual freedom and equality of opportunity are available only to those who control the means of production, distribution, and exchange. 'Control' in this sense means owned or regulated collectively—either by the state or by the community—to promote social equality and reduce economic disparities. Socialism is understood differently by various theorists and political movements, but it generally encompasses ideas critical of capitalism and visions of a future society intended to replace it. Three broad categories can be identified: (1) state socialism, characterized by a state-owned, centrally administered economy controlled by a dominant communist party that mobilizes the population to create a classless society based on Marxism-Leninism; (2) market socialism, incorporating market mechanisms within a socialist framework to balance efficiency with social ownership; and it involves a non-capitalist form of investment and production that is planned, along with a market form of distribution; and (3) democratic socialism, which emphasizes the balance of power between workers and capitalists, aiming to abolish private property in the economic sphere and ensure production is controlled democratically by the workers in conjunction with a workers' state (Bolton, 2020).

Socialist systems divide into non-market and market forms—both supporting economic and social interventions to promote social justice. The core principles of socialism include public ownership of industries and services, wealth redistribution (through progressive taxation and social programs), social welfare (e.g., providing universal access to basic needs like healthcare and education), and economic planning to prioritize societal well-being (reducing inequality) over profit maximization. Socialism's evolution in the twenty-first century has led to, what some call, 'new socialism' (Cockshott & Cottrell, 2000; Lebowitz, 2016). New socialism refers to modern adaptations of socialist principles that integrate elements of digital technology, sustainability, and decentralized governance. While interpretations vary, some key aspects of new socialism include:

- *Digital and platform socialism*: Utilizing technology and AI to distribute resources more equitably, often through decentralized networks rather than traditional state control.
- *Green socialism*: Prioritizing sustainability, climate action, and collective ownership of renewable energy infrastructure.

- *Hybrid models*: Blending socialist policies (e.g., reducing inequality) with aspects of capitalism, such as state-managed market economies seen in China.
- *Universal basic services*: Expanding state-provided healthcare, education, and housing to ensure social welfare without necessarily abolishing private enterprise.

Contemporary or new socialism represents an evolution of socialist ideas adapted for the twenty-first century, often incorporating digital governance, ecological concerns, and reformed economic models to address contemporary global challenges. Given that socialism is a system built on social ownership and the planned coordination of the economy, now in the twenty-first century it seems to be moving closer to stakeholder capitalism, which emphasises that all those who have a stake in the economy can influence decision-making (Schwab & Vanham, 2021b). It involves socialized means of production allowing for collectively rational social control over production, distribution and development, as well as democratic control over these processes. This is socialism, but the same principle can be applied to stakeholder capitalism (i.e., collective responsibility). If one also considers that democratic socialism, a branch of socialism, advocates for a democratically managed economy and society, ensuring that resources and decision-making serve the needs of the entire population rather than a privileged few (e.g., Nordic countries), it seems very similar to stakeholder capitalism. Under market socialism, democratic governance is combined with socialist economic policies. As such, socialism does not require full government control; instead, businesses and institutions should be operated by the workers and consumers directly impacted by them (their stakeholders). This can take the form of worker-owned cooperatives or publicly owned enterprises governed by both employees and consumer representatives (Bolton, 2020). Again, this is quite similar to stakeholder capitalism, especially given that new socialism involves reclaiming socialism as a concept with an emphasis on cooperation and humanism, rather than competition.

New socialism presents a vision for a more just, democratic, and efficient society that learns from the successes and failures of past socialist experiences. It is a 'humanist' type of socialism. New socialism prioritizes human beings over machines or the state, emphasizing the development of human capacity and individuality as the real wealth of society. It aims to create a society where objective wealth is used to fulfil the worker's need for development, allowing for the "all-round development of the individual" (Lebowitz, 2013). Social ownership of the means of production is crucial to ensure that communal productivity is directed toward the free development of all, not just the private goals of capitalists or state bureaucrats. Social production focuses on satisfying social needs and is part of an organic system. New socialism advocates for new systems built on cooperation. It recognizes the need to reinvent socialism, moving away from the expansion of the means of production or state direction. It emphasizes the 'triangle' of social property, social production, and the satisfaction of social needs as an interconnected system.

The state's role in new socialism is envisioned as society's "own living forces," subordinate to society rather than standing above it (Lane, 2009: 102). This involves fostering change in both circumstances and human beings, rejecting capitalist measures of accounting and efficiency in favour of socialist accounting that recognizes human development. New socialism aims to replace private ownership with social ownership, worker management with workplace autocracy, and productive activity based on individual self-interest with solidarity and community. It focuses on subordinating all elements of the old society to build a new one. This corresponds with Klaus Schwab's vision of building a new society through the 4IR in which stakeholder capitalism becomes "the third system" (to state capitalism and shareholder capitalism), "overcoming much of their shortcomings" and being "the preferred economic system, the one we ought to implement going forward" (Schwab & Vanham, 2021b). What is quite ironic from a socialist perspective is that in both shareholder and state capitalism, the dominance of one stakeholder over the others is the system's primary weakness. Shareholder capitalism prioritizes shareholder interests above all else, while state capitalism concentrates excessive power in the hands of the government.

State capitalism blends government intervention with market mechanisms to achieve economic and political goals. In the global context, the rise of state capitalism and the expansion of China are reshaping the vision of development, with new socialism responding to these changes (Alami et al., 2021). On the question of whether state capitalism is similar to new socialism, the answer depends on how one interprets the goals and mechanisms of each system. Table 2 draws a comparison that provides perspective.

As Table 2 points out, both systems involve significant state intervention, but their objectives differ—state capitalism focuses on national power and economic growth, while new socialism emphasizes social equity. Both systems involve state ownership, but new socialism typically extends this to a broader range of industries and services. Furthermore, state capitalism retains markets, while new socialism seeks to minimize their role. Both systems critique unfettered capitalism, but state capitalism does not inherently seek to dismantle capitalist structures. The same is true for stakeholder capitalism in that it seeks a new way to reform capitalism, making it more sustainable and inclusive. What is interesting in Table 2 is that there seems to be more similarities between new socialism and stakeholder capitalism, than with state capitalism. The latter seems to be more 'socialistic' than new socialism, which is quite ironic.

State capitalism is not the new socialism, but it does blur the traditional boundaries between capitalism and socialism in significant ways. While socialism prioritizes collective ownership, wealth redistribution, and public welfare, state capitalism operates within a market-driven framework, but with strong government control over key industries and economic strategies (Bremmer, 2009). State capitalism is not a new form of socialism, but rather a hybrid model that combines government intervention with capitalist competition. It challenges traditional free-market capitalism while avoiding the complete economic restructuring associated with socialism. Yet, why is there confusion about the blurring of the boundary

Table 2 Comparing new socialism, state capitalism and stakeholder capitalism

Aspect	New socialism	State capitalism	Stakeholder capitalism
Core principle	Human-centred, prioritizing human development, cooperation, and community. Advocates for collective ownership, democratic governance, and equitable distribution of resources	State plays a strong role in organizing the economy, supervising capital accumulation, or directly owning/controlling capital (key industries), while allowing some private enterprise. Markets are used primarily for political gain	Balances the interests of all stakeholders (employees, customers, shareholders, communities, and the environment) with financial performance, competitiveness, and inclusive economic growth
Role of the state	Strong state involvement in planning and regulation, often aiming for the state to be subordinate to society rather than standing above it	The state plays a dominant role in organizing the economy, managing major industries for national goals, directs national development, intervenes in markets, and distributes wealth and opportunities	The government does not supplant the board as the ultimate decision-maker. The state acts as a regulator, ensuring businesses consider broader societal and environmental impacts
Ownership of enterprises	Social ownership of the means of production is crucial, ensuring communal productivity is directed toward the free development of all. Can involve public, cooperative, or decentralized worker ownership	State can have various levels of ownership, from wholly-owned enterprises to minority equity positions. SOEs dominate key sectors, with some private ownership	Private ownership is generally maintained, but with a greater emphasis on the responsibilities of owners to stakeholders. Businesses operate with a focus on long-term value for all stakeholders
Role of the market	A mix of market mechanisms and collective ownership with social welfare priorities. Social production focuses on satisfying social needs and is part of an organic system. Focused on reforming or remaking the market	Market-driven but heavily controlled by the state, often through SOEs and strategic planning. State uses markets for political gain. Market mechanisms are used for the promotion of geoeconomic and geopolitical goals	Operates within a free-market system but seeks to redirect capitalism to deliver on goals related to inclusion and sustainability. Corporate decisions integrate social and environmental responsibilities
Profit orientation	Profit is secondary to social well-being and sustainability	Profit is pursued, but within a framework of state control and national strategic interests	Profitability is important but balanced with social, environmental, and ethical considerations

(continued)

Table 2 (continued)

Aspect	New socialism	State capitalism	Stakeholder capitalism
Decision-making power	Decentralized and democratic, with worker and community participation	Centralized in government or ruling elites	Shared among shareholders, employees, communities, and other stakeholders
Relationship to capitalism	Aims to replace key aspects of capitalism with socialist alternatives, subordinating all elements of the old society to build a new one	Can be seen as a variant of capitalism where the state plays a more dominant role. May compete with other forms of capitalism	Is framed as a counterpoint to shareholder capitalism, seeking to reform and redirect existing capitalist systems
Ultimate goals	Development of human capacity and individuality. Satisfaction of social needs and building a truly human community	The state functions as the leading economic actor, often with goals related to national development, geopolitical influence, or maintaining political control	Achieve sustainable development goals and address challenges such as climate change, inequality, and digital exclusion
Global examples	Emerging models in cooperative economies, digital socialism, and some Nordic influences	China, Russia, Gulf States, Vietnam, Singapore, and other countries with significant state involvement in the economy	European social market economies like Germany, some US corporations (e.g., ESG-focused firms), Japan, and WEF promotes it strongly

Sources Lebowitz (2016), Alami et al., (2021), Bremmer (2009), Samans and Nelson (2022)

lines? One reason is that the failures of unfettered capitalism (e.g., inequality, unsustainability and financial crises) have renewed interest in state intervention, whether framed as socialism or state capitalism. More reasons for why state capitalism resembles socialism but is not the same, include:

- *Government control without full socialization*: Countries like China, Russia, and even some Western economies with strong state intervention (e.g., industrial policy in the EU) use state capitalism to drive economic growth. However, they do not fully socialize wealth or eliminate private enterprise.
- *Market integration*: Unlike socialist economies that may reject capitalism, state capitalist systems rely on global markets and private-sector innovation to compete internationally.
- *Economic efficiency over equality*: State capitalism often prioritizes national economic strength and competitiveness rather than broad wealth redistribution.

While state capitalism is not the new socialism, it shares key similarities in its reliance on state intervention. Whereas state capitalism focuses on national economic strength and strategic control, new socialism aims to dismantle capitalist structures to achieve social equity (Cockshott & Cottrell, 2000). However, in a world where traditional socialism has waned and neoliberalism faces criticism, state capitalism has emerged as a pragmatic alternative that blends state control with market mechanisms. Whether it represents a new form of socialism depends on how one defines socialism and the extent to which state capitalism addresses social and economic inequalities. Historically, some states that have identified as 'socialist' have, in practice, operated more like state capitalist systems. This has led to debates about whether these systems truly represent socialism.

Hence, perhaps the main cause of the confusion is how China blurs the traditional boundaries between capitalism and socialism through its unique model of state-led capitalism within a socialist political structure—that is, by combining market-driven economic policies with strong state intervention and centralized political control. While state capitalism typically refers to a system where the government plays a dominant role in the economy while allowing some private enterprise, socialism traditionally emphasizes collective ownership and wealth redistribution. China's model blends these two by maintaining a socialist political framework while operating within a highly capitalist global economy. This hybrid system, often called 'socialism with Chinese characteristics', combines market-driven economic growth with strong state control, challenging conventional definitions of both capitalism and socialism (Boer, 2021). This is how China blends the two models:

- *Hybrid ownership model*: While SOEs control strategic industries, the private sector drives much of China's innovation and economic growth, contributing over 60% of GDP. The government uses 'mixed ownership reforms' to integrate private capital into state-led enterprises.
- *Authoritarian control in a market economy*: Unlike Western capitalist democracies as well as traditional state capitalism (e.g., in Russia or Gulf states), China allows capitalist-style competition within a socialist framework. The Communist Party ensures that even private firms align with national interests, using regulatory crackdowns (e.g., on technology and education sectors).
- *Common prosperity and wealth redistribution*: While capitalism thrives in China, recent policies under Xi Jinping's 'Common Prosperity' agenda show a socialist tilt, aiming to reduce wealth inequality through taxation and corporate social responsibility (Wu, 2022). It signals a shift from extreme wealth concentration to wealth redistribution, increasing regulation of big tech and private capital.
- *Global economic integration with political independence*: China uses capitalist tools (global trade and foreign investment) but rejects Western liberal values, asserting state control over capital.
- *Technology and surveillance as a state asset*: China merges capitalist innovation with socialist governance, using tech advancements to strengthen state control,

e.g., the social credit system[3] and AI-driven governance: predictive policing and surveillance tools (like AI-powered cameras) and big data to monitor and maintain discipline over its population (e.g., internet censorship (Great Firewall), blocking Google, Facebook, and X).

China's model challenges the idea that capitalism and socialism are mutually exclusive (that a country is either fully state capitalist or purely socialist). It demonstrates that a state can embrace market mechanisms for economic efficiency while maintaining strict political and ideological control. This is mainly due to China's unique path of development, which combines elements of both systems (Lane, 2013). Early in its reform process, China started practicing market socialism, layering market activity on top of a planned economy. Agricultural reforms, encouragement of private enterprises, and special economic zones were all features of this stage. China's economic reforms represent an adaptation to the capitalist world economic order, moving away from the Soviet paradigm of development and Maoism. This adaptation involves sinicizing Western market capitalism with 'Chinese characteristics' (Xing & Shaw, 2013). The lines between state and private sectors in China are blurred, with the state retaining significant influence over nominally private firms. This is achieved through mechanisms such as party membership within enterprises, state approval for investments, and the intertwining of political and economic objectives.

China's party-state has demonstrated a resilient capacity to sinicize[4] and internalize foreign ideas and practices, adapting them to its unique political culture and rationality. China's model challenges the American model of liberal values and political pluralism by evolving a national government that legitimizes a state-led economic order. The state sector responds to both economic incentives and state policies, guiding the overall direction of the economy, including the private sector (Gu et al., 2016). Even with a growing private sector, the state emphasizes the dominance of public ownership in its economic system. China has evolved a hybrid system with a strong political faction seeking further integration into the global neoliberal economy while maintaining government control. This involves navigating the tensions between globalizing parent companies and nationally based regional companies. This hybrid system influences debates about the future of capitalism, particularly in areas like state intervention, industrial policy, and digital governance. Particularly, debates abound about whether China's model represents

[3] China uses its social credit system to assess the 'creditworthiness' and 'trustworthiness' of people and businesses by collecting digital data on their behavior. To reward compliance and punish rule-breaking, this system uses data on social and economic activities to create a 'social credit score' of individuals and businesses, affecting access to discounts, school admissions, loan interest rates, etc. It flags dissent and suppresses politically sensitive content.

[4] Sinicizing (or sinification) is making something more Chinese in character, culture, or influence. It involves adapting or assimilating non-Chinese people, institutions, or practices into Chinese traditions, governance, or ideology.

a genuine alternative to neoliberal capitalism or simply a different form of state-managed capitalism, thus influencing global discussions on governance, economic planning, and the role of the state in capitalism. What makes it even more significant is Klaus Schwab's view—as an influential previous leader of the World Economic Forum—that China is a "role model" for the future (Billingsley, 2022). Such views from the West further blurs the boundary lines between socialism and capitalism.

Lastly, what is interesting though is how comparatively similar new socialism and stakeholder capitalism is. This can even be observed in Table 2. These similarities stem from their aims to address the shortcomings of neoliberal capitalism and to broaden the focus of economic and corporate activities beyond shareholder value and profit maximization (Mazzucato, 2024). While they emerge from different ideological traditions, both frameworks advocate for a more inclusive and equitable economic system. Both concepts share common ground in the following aspects:

- *Critique of shareholder primacy*: Both stakeholder capitalism and new socialism challenge the single-minded, conventional focus on shareholder value maximization. They broaden the corporate purpose to include creating long-term value for all key stakeholders and advocates for a new social contract among businesses, labour, and the state.
- *Emphasis on equitable distribution of value*: Both concepts prioritize a more equitable distribution of value. Both aim to create a fairer economic system that serves broader society rather than just elites (Samans & Nelson, 2022). Both reject pure free-market capitalism in favour of more inclusive economic structures.
- *Role of state institutions*: New financial institutions, such as state investment banks, can provide patient capital to incubate innovation ecosystems and ensure a more equitable distribution of value. They acknowledge that governments can embed the notion of shared value creation in their corporate governance statutes and cultures.
- *Stakeholder cooperation*: Both value principles such as stakeholder cooperation and engagement. New socialism, through democratic socialism, seeks to reconcile the interests of worker-controlled firms with those of society as a whole (Mhlanga, 2022). Both advocate for giving workers more voice in the economy (worker empowerment).
- *Systemic approach*: Both concepts address systemic issues, such as climate change, inequality, and other challenges. They incorporate sustainability as a key principle. Stakeholder capitalism is a response to systemic risks and aims to balance the interests of various stakeholders (Beck & Ferasso, 2023). New socialism revisits the structures and decisions that shape capitalist systems to address multiple crises.

New socialism leans more towards state involvement and redistribution, while stakeholder capitalism retains a market-driven approach with corporate responsibility mechanisms. Both advocate for a broader consideration of social and environmental impacts. However, while new socialism often calls for systemic transformation, stakeholder capitalism seeks to reform existing capitalist structures to better serve society as a whole. Both are progressive in nature and it would not be a surprise if—as they continue their trajectories—they keep blurring the lines between capitalism and socialism. Progressive thinking from both the West and the East seems to stimulate such a merge, as modern economic and political models incorporate elements of both systems. This blending or synergy is evident through trends mentioned, as well as, expanding the role of the state in capitalism (e.g., the Green New Deal), social safety nets in market economies, corporate responsibility, worker participation in economic decision-making, and surveillance capitalism. These trends suggest that 21st-century economies are evolving beyond the capitalism vs. socialism binary, creating a mixed model that blends market efficiency with social protections.

4 Synergy and the SDGs?

In view of Agenda 2030, a vital question is how the progressive synergy or blending between capitalism and socialism is affecting the achievement of the SDGs. Different aspects of traditional capitalism and socialism—positive and negative—have an impact on SDG-attainment. In the case of capitalism, private ownership, free markets, and profit maximization play a vital role in, for example, creating decent work and economic growth (SDG 8) and other SDGs. While capitalism focuses on driving innovation and economic growth, it may also lead to income inequality and environmental exploitation, which presents a challenge to reducing inequality (SDG 9) and getting to zero poverty (SDG 1), and creating sustainable cities and communities (SDG 11). Socialism prioritizes social ownership, equitable distribution, and government intervention, which might help to address the very negative impacts that capitalism has on SDG-attainment. Socialism also aims to provide social safety nets and reduce inequality, but can sometimes stifle innovation and economic efficiency, which limits SDG-attainment. What is significant is that new concepts like stakeholder capitalism and new socialism are expanding the possible (and positive) implications for the achievement of the SDGs. Both frameworks emphasize social equity, environmental sustainability, and inclusive economic growth, which align closely with the SDGs.

It appears that the blending of socialism and capitalism—through progressive policies, stakeholder capitalism, and state intervention—may indeed have a significant impact on achieving the SDGs by addressing economic, social, and environmental challenges in a more balanced and inclusive way (Brundenius, 2020). The interplay between socialism and capitalism with regards to their synergy, and especially their more progressive versions, can have both positive and negative impacts on SDG-attainment. Firstly, potential benefits of blending capitalism and socialism for achieving SDGs include:

- *Sustainable innovation*: Government regulation (socialist influence) can steer capitalist innovation towards sustainable technologies and practices (SDGs like clean energy, responsible consumption).
- *Peaceful competition*: A peaceful competition between the US and China, with a focus on saving the planet for future generations, might allow for a variety of social systems that mix elements of capitalism and socialism in new ways (Sutterlütti & Meretz, 2023). This could create more room for achieving the SDGs.
- *Inspiration for socialist strategies*: The UN's SDG agenda can serve as a source of inspiration for strategies for new socialism, addressing many issues that such strategies need to deal with. Many of the SDG goals and targets align well with traditional socialist aims, such as ending poverty and hunger, improving living conditions, ensuring inclusive access to education, and reducing inequality.
- *Social safety nets*: Socialist principles can ensure basic needs (SDGs like no poverty, zero hunger and good health) are met, while capitalist mechanisms drive economic activity.
- *Broader perspective*: A blend may incorporate an Anthropocene perspective (humanity having an environmental impact on the earth) on development, which is implicitly present in the SDGs, including sustainable consumption and production, use of natural resources, and addressing climate change (Baer & Singer, 2023).
- *Stakeholder capitalism*: This evolving model, which blends aspects of both systems, encourages businesses to consider social and environmental impacts alongside profits, aligning with the SDGs. Stakeholder capitalism, which integrates ESG issues, is seen as critical for achieving the SDGs (Beck & Ferasso, 2023). It can foster sustainable development patterns at all levels and help achieve SDGs related to industry, innovation, infrastructure, and partnerships. Stakeholder cooperation and engagement are elementary for fostering sustainable development and contributing to the achievement of the SDGs.

Secondly, potential drawbacks and challenges of blending capitalism and socialism may stifle progress toward the SDGs. These include:

- *Contradictions of capitalism*: Critics highlight that the SDGs may not sufficiently emphasize planetary boundaries or transform the capitalist global system. The SDGs may be compromised by the contradictions of capitalism (Vu & Long, 2023). While it might stimulate growth and even improve GDP per capita, unregulated capitalism can exacerbate income inequality (SDG 10), hindering progress on other SDGs.
- *Implementation difficulties*: Even with goal-setting, implementation is a significant challenge, hinging on institutional factors such as translating global goals to national ones and the extent of national commitment. Furthermore, balancing

socialist principles with capitalist structures can be complex, requiring careful policy design and governance.

- *National ownership challenges*: National ownership of the SDGs can be intertwined with critiques of the SDGs. If the SDGs offer a positive and transformative vision, national ownership might pose a threat to transformation (Pierleoni, 2020). Achieving the SDGs also requires global cooperation, which can be hindered by geopolitical tensions and differing economic systems, resulting in regional blocks and global alliances based on shared national interests.
- *Greenwashing*: Businesses might exploit the SDG framework to 'social wash' their capitalist motivations, selectively pursuing goals while circumventing others. The fact is, the pursuit of profit in a purely capitalist system may lead to unsustainable resource use and environmental damage (affecting SDGs like climate action and life below water).
- *Superficial consideration of capitalism*: Some studies suggest there are structural limitations to achieving the SDGs but do not clearly define or theorize these structures, with a general omission of capitalism and class relations (Baer & Singer, 2023; Brundenius, 2020; Hornung et al., 2022).
- *Power structures*: The SDGs may exist as part of a superstructure at the global level that furthers the subordination and exploitation of developing countries. In addition, powerful interests (e.g., corporations and wealthy elites) may resist changes that reduce their profits or influence.
- *Market integration issues*: Market integration and market forces may contradict goals related to biodiversity conservation (SDG 15) and climate action (SDG 13). In some socialist models, there could also be a lack of investment into innovation that could help solve SDG-related problems.

Overall, the synergy between socialism and capitalism in the context of the SDGs is complex. While elements of both systems can contribute to certain goals, inherent contradictions and challenges in implementation and power dynamics need to be addressed. The blending of these systems manifests differently in various countries, leading to diverse outcomes. Achieving the SDGs requires a balanced approach that leverages the strengths of both systems. Effective governance, strong regulations, and social responsibility are crucial for aligning economic activity with sustainable development. In essence, the synergy between socialism and capitalism can either accelerate or hinder SDG progress, depending on how effectively it is managed. On the question of which specific SDGs are most affected by the convergence between the two systems, they are identified in Table 3.

The synergy between socialism and capitalism, as seen in progressive approaches like stakeholder capitalism and new socialism, can create a more equitable, sustainable, innovation-driven, and inclusive economic system that aligns closely with the SDGs. By combining the strengths of both frameworks—such as socialism's focus on equity and capitalism's emphasis on innovation and efficiency—this blended approach can accelerate SDG progress. The potentially positive outcomes of such a balanced model for SDGs include: (1) ensuring wealth distribution while fostering economic dynamism and growth; (2) balancing state

Table 3 SDGs influenced by the synergy between capitalism and socialism

SDGs	Capitalist aspects	Socialist aspects	Blending effect
SDG 1 & 8 (combined)	Market-driven innovation and private sector investment create jobs and economic opportunities	Stronger social safety nets (e.g., universal basic income and welfare programs) help reduce poverty and economic inequality	Progressive taxation and impact-driven capitalism ensure that economic growth benefits all rather than just a wealthy elite
SDG 3 & 4 (combined)	Private investment in edtech, biotech, and healthcare innovation enhances service quality	Free or subsidized healthcare and education systems improve access for marginalized groups	A hybrid model ensures universal access while still encouraging competition-driven improvements
SDG 8	Gig economy and remote work create flexibility and opportunities for entrepreneurship	Stronger labour protections, union rights, and co-determination models (e.g., worker-owned enterprises)	Regulated capitalism ensures fair wages, ethical supply chains, and workers' well-being without stifling business growth
SDG 9	Venture capital, tech start-ups, and private sector R&D push innovation forward	Government funding in critical infrastructure and research (e.g., public transit, AI research and energy grids)	State investment de-risks innovation, making breakthrough technologies (e.g., AI and clean energy) more accessible
SDG 10	Entrepreneurship and access to markets provide economic mobility	Labour rights, progressive wages, and redistributive policies (e.g., universal healthcare and education) reduce wealth gaps	A mixed economy allows both wealth creation and redistribution, ensuring inclusive prosperity
SDG 12 & 13 (combined)	Green technology, circular entrepreneurship, and ESG investing promote eco-friendly business models	Government-led climate policies (e.g., carbon taxes and green subsidies) enforce sustainability	Public–private partnerships drive climate action by combining state regulation with market innovation

Sources Guttmann (2023), Mhlanga (2022), Millán-Acevedo and Gómez-Bruna (2025)

intervention with market efficiency; (3) developing sustainable business models and communities; and (4) prioritizing human well-being alongside profit. However, it is important to note that successful implementation will require overcoming significant political, economic, and institutional challenges. This is where progressive thinking plays a vital role. Lastly, a word of caution also for the risks of exploitation (by capitalism) and the centralization of power (by socialism) in case unsuccessful synergy leads to backlash.

5 What About the Resurgence of Trumpian Capitalism from 2025?

What is unmistakable about Donald Trump is his 'naked capitalism' stance. US President Trump is a true capitalist—the most authentic and the most experienced who has ever been in the White House. Nothing is more fundamental to capitalism than what its proponents refer to as self-interest—what others might simply call selfishness—the "I-got-mine-you-get-yours" mentality (Nimtz, 2019). Could there be a more accurate portrayal of Trump's conduct? His defining trait is an overwhelming self-focus. This is where the 'America First' policy originates from, though not without concern for the rest of the world, hence Trump's approach of 'Peace Through Strength'. The 'Trumpian capitalism' on the rise is unbridled rather than restrained. In this respect, Trump is a Republican in the classic mould. For the vast majority of the progressive left, including those seeking to reform capitalism, Trump and what he represents is an inconvenient truth. Yet, he is in the office of what is called "the most powerful person on earth" (Khalili, 2023).

Due to growing competition from the East—especially China—a new change in the dynamics of capitalism's evolution is taking shape. The Trump administration's approach to economic growth and progress leans much more towards conventional shareholder capitalism, which is different than the past decade's progressive trajectory in capitalism's evolution. Being much more sensitive to global competition than the previous administration, the new Trump administration is on the charge to reassert US dominance in the world economy. Being a business man himself, Donald Trump wants to 'Make America Great Again' through economic success, which is primarily orientated towards classic shareholder capitalism. The drive for higher profits, maximizing shareholder value and rapid economic growth at any cost may now become the new norm of capitalism, which strongly redirects the previous trajectory. Given America and the dollar's influence in the world economy and Donald Trump's unique approach, key questions to consider include: how will this happen and what are the implications for Western capitalism in the broader sense—especially Europe's role in it and its possible repositioning—i.e., will Europe still remain a key ally to the US and/or will it remain committed to capitalism's more progressive trajectory?

First, what exactly is 'Trumpian capitalism'? It refers to an economic and political approach associated with US President Donald Trump, characterized by a mix of economic nationalism, deregulation, protectionism, and corporate favouritism. In some ways, it represents a departure from traditional free-market capitalism and incorporates populist rhetoric with policies favoring businesses, especially large corporations and industries perceived as strategically important. It entails a strong focus on domestic industries, protectionist trade policies to gain a competitive advantage (e.g., tariffs), and a rejection of globalist economic frameworks. Based on the economic policies and practices of Trump's first term in office (2017–2021) and the way he started his second term from 2025 onward, the following features and principles of Trumpian capitalism can be identified:

- *Economic nationalism and protectionism*: Trumpian capitalism prioritizes the growth and protection of domestic industries, particularly manufacturing, energy, and agriculture. This is often framed as 'putting America first'. Policies are aimed at stimulating domestic production and employment, such as encouraging government agencies and businesses to purchase American-made goods and hire American workers (Fuchs, 2017). It entails trade wars and tariffs, particularly against China, to reduce reliance on foreign manufacturing and reduce trade deficits. Trump's policies tend to also impose tariffs on companies that outsource production but sell products back in the US. His approach like, for instance, the huge tariff hikes on most countries in 2025, stands in opposition to longstanding global economic agreements, such as when he pulled the US out of the Trans-Pacific Partnership (TPP) and now also The North American Free Trade Agreement (NAFTA).
- *Deregulation*: Trumpian capitalism emphasizes reducing regulatory burdens on businesses, particularly in industries like energy, finance, and manufacturing. This is seen as a way to stimulate economic growth and job creation. The Trump administration rolled back numerous environmental regulations, such as those related to climate change, clean air, and water protections, to support industries like coal, oil, and gas.
- *Tax cuts and corporate-friendly policies*: A major legislative achievement of the first Trump administration was the 2017 Tax Cuts and Jobs Act. This law significantly reduced corporate tax rates and provided tax cuts for individuals, with the aim of stimulating investment and economic growth (Mollan & Geesin, 2020). Now this is done through the 'One Big Beautiful Bill'. Trumpian capitalism favours large corporations (especially in industries like fossil fuels, real estate, and Wall Street) and corporate leaders, with policies designed to boost corporate profits, stock market performance and short-term economic growth, even at the expense of long-term sustainability and equity.
- *Populist rhetoric*: Trump's economic policies are often framed as benefiting the 'forgotten' working class, particularly in regions affected by deindustrialization and globalization. He appeals to working-class voters through rhetoric about bringing back jobs, especially in manufacturing and coal.
- *Scepticism of globalism*: Trump withdrew from or criticized international agreements and organizations, such as the Paris Climate Agreement, the World Health Organization (WHO), and the WTO, reflecting a broader scepticism of globalist economic and political frameworks (McGaughey, 2018). Given his scepticism of international free trade agreements, Trump prefers bilateral trade deals over multilateral agreements, arguing that they allowed the US to negotiate more favourable terms.

Trumpian capitalism should be viewed through the lens of 'Trumpism' or 'Trumpology,' described as Trump's style of ideology and a way of thought encompassing hyper-individualism, hard labour, leadership, a friend/enemy scheme, and even a form of Social Darwinism.[5] His leadership style suits those who would break the hold of a faceless plutocratic class in Washington. Trumpian capitalism is a hybrid of protectionist state capitalism and corporate-friendly neoliberalism, emphasizing economic nationalism, deregulation, and short-term market gains while using populist rhetoric to appeal to working-class voters. It blurs the line between protectionism and neoliberalism, using nationalist rhetoric while enacting policies that favour big business and financial markets. It challenges global capitalism while reinforcing corporate power, making it distinct from both traditional shareholder capitalism and emerging stakeholder capitalism. Table 4 provides a comparative summary of the differences between these types of capitalism and Trumpian capitalism, as well as how the latter differs from state capitalism. While traditional (neoliberal) capitalism emphasizes free markets, globalization, and minimal government intervention, Trumpian capitalism diverges in several key ways as seen below.

Trumpian capitalism's unique blend is also associated with authoritarian statism. Both emphasize the concentration of power, nationalism, and a departure from traditional democratic norms in economic governance. More specifically, Trumpian capitalism favours a strong, interventionist executive who overrides traditional checks and balances to shape economic policies. Likewise, authoritarian statism, as theorized by Poulantzas and Hall (1978), describes a system where the state consolidates authority, weakening democratic institutions while maintaining capitalist structures. Trumpian capitalism promotes protectionist policies, such as tariffs and trade wars, as a way to favour domestic industries. Similarly, authoritarian statism relies on state control over markets, often reducing foreign influence while maintaining corporate interests.

Akin to Trumps corporate favoritism, authoritarian statism creates an elite economic structure where state-backed firms and powerful actors dominate. Similar to Trump's deregulation (weakening oversight on businesses), authoritarian statism maintains capitalism but erodes institutional autonomy, ensuring the state directs economic activity to serve political ends. The main concern with this association is that economic nationalism, crony capitalism, and executive overreach can erode democratic institutions and concentrate power in the hands of the state and corporate elites. In a world of uncontrolled technological advancement, this concern is heightened in the West, given influential scholars like Shoshana Zuboff's (2019: 8) claim that "surveillance capitalism is an assault on human autonomy".

This leads to the question as to whether Trumpian capitalism is the new form of authoritarian capitalism. The latter refers to an economic system where the state plays a dominant role in guiding the economy, often under an authoritarian political regime (Bloom, 2023). In view of the link between the Trump presidency

[5] Social Dawinism is a set of ideologies that apply the biological concepts of natural selection and 'survival of the fittest' to human societies. In the Trumpian context, he is doing everything possible to gain a competitive advantage.

Table 4 Comparing Trumpian capitalism with other capitalist models

Aspect	Trumpian capitalism	Neoliberal capitalism	Stakeholder capitalism	State capitalism
Core philosophy	Economic nationalism, crony capitalism, and elite-driven growth	Free markets, deregulation, and minimal state intervention	Balancing profit with social responsibility and long-term sustainability	Government plays a dominant role in the economy, often owning or controlling key industries
Role of the state	Strong executive power influencing markets; favours protectionism	Minimal state intervention, emphasizing privatization and deregulation	Regulatory role to ensure businesses serve multiple stakeholders	State actively participates in and directs the economy
Key economic policies	Protectionism, tariffs, tax cuts for the wealthy, deregu-lation, and corporate favouritism	Deregulation, privatization, free trade, and tax cuts to incentivize investment	ESG policies, corporate responsibility, and inclusive economic growth	State ownership of enterprises, strategic industrial planning, and economic intervention
Wealth distribution	Benefits primarily the wealthy elite and corporations; trickle-down economics	Market-driven inequality; wealth accumulates at the top but with economic mobility	Equitable wealth distribution through corporate accounta-bility and sustainability measures	Often fosters income disparity between politically connected elites and the general population
Labour relations	Pro-business policies, weak labour protections, emphasis on domestic job creation via tax incentives	Weak labour protections, favouring market-determined wages and employment	Strengthens labour rights, promotes fair wages, and considers workers as stakeholders	State-influenced labour policies, sometimes suppressing unions but ensuring employment stability
Trade approach	Protectionist, often anti-globalist, favours bilateral over multilateral trade agreements	Globalist, pro-free trade, supports international markets and institutions like the WTO	Encourages responsible globalization with fair-trade policies and ethical supply chains	Uses trade strategically, often blending open-market policies with protectionist state intervention

(continued)

Table 4 (continued)

Aspect	Trumpian capitalism	Neoliberal capitalism	Stakeholder capitalism	State capitalism
Corporate power	Strengthens ties between government and select corporations (crony capitalism)	Corporations operate independently with minimal regulation	Corporations held accountable for broader societal impact, not just shareholder returns	SOEs and national champions dominate key industries
Democratic implications	Weakens institutions through executive overreach and populist rhetoric	Maintains formal democratic structures but often favours corporate lobbying power	Seeks a balance between democracy and corporate influence to serve all stakeholders	Can coexist with democracy but often leans toward authoritarian economic control
Market regulation	Selective deregulation that benefits allies and large corporations	Minimal regulation to promote free-market competition	Enforces regulations that promote ethical business and sustainability	Heavy state intervention, controlling strategic industries and economic direction
Key examples	Trump-eras of US economic policies (2017–2021 and 2025–2029)	Reaganomics, Thatcherism, and post-1980s globalization	European social market economies, Business Roundtable statement of 2019	China, Russia, Gulf States, and Singapore

Sources Huang (2022), Jaffee (2019), Mollan and Geesin (2020), Mhlanga (2022)

and the rise of right-wing populism, Trumpian capitalism does exhibit several features of authoritarian capitalism. It represents a hybrid model that incorporates elements of 'Trumpology', such as top-down leadership, economic nationalism, the friend/enemy scheme, authoritarian statism, and Social Darwinism within a nominally democratic system. That said, Trumpian capitalism lacks the authoritarian governance and restrictions on civil liberties that define authoritarian capitalist systems. It seems to be true that tendencies of Trump's rule could advance authoritarian statism in US capitalism. However, the criteria for judging this development include the decline of the rule of law, the concentration of power, and the limitation of political opposition to enforce policies supporting the capitalist class. Differences in governance, political structure, and the role of democracy exist because Trumpian capitalism does not involve the suppression of political freedoms or centralized authoritarian control. It also does not exhibit full state control over the economy. While Trumpian capitalism shares some traits with authoritarian capitalism, it is not a form of authoritarian capitalism—but may be a step toward it. Trumpian capitalism is better understood as a populist, nationalist, and protectionist variant of capitalism that weakens democratic norms and diverges from traditional free-market and neoliberal models, but stops short of authoritarianism.

The main concern, though, is that the Trump administration would deepen features of the neoliberal political-economic arrangement and make it less likely that this arrangement can be challenged through political forms of democratic expression. For progressives, the main concern is that Trumpian capitalism might drive a stake through stakeholder capitalism.

There is no doubt that Trumpian capitalism has become a disruptive force in the global economy, especially since his second term of office started in 2025. All his policies are setting the progressive agenda back, almost as if he and his dynamic cabinet revengefully wants to make up for the four 'lost years' under Joe Biden. By putting 'America first' more aggressively than ever, Trump is putting the world on notice. After blaming and warning Canada, Mexico and China about the influx of fentanyl at the US borders, he initially decided to impose a blanket 25% tariffs on imports from Canada and Mexico. Trump then also doubled the tariff on all Chinese imports to 20% from 10% (Buchwald, 2025). China and Canada swiftly responded with tariffs on US goods, raising the risk of a harmful trade war. Mexico followed suit by imposing its own tariffs on American goods. All these developments culminated in, what Trump called, America's "Liberation Day", on 2 April 2025. After raising extreme tariffs on countries, Trump invited 'the world' to come and negotiate or renegotiate trade agreements with the US. Since then, many countries have come to the Oval Office to agree on new terms with regards to trade and investment in the US.

Former Canadian Prime Minister Justin Trudeau warned in a press conference on 4 March 2025 that Canada "will not back down from a fight". He further said, "We will also be challenging these illegal actions by filing dispute resolution claims at the WTO" (Norman, 2025). That was the end of his reign, which was followed by Mark Carney becoming the new Prime Minister of Canada.

Lin Jian, a spokesperson for China's Foreign Ministry, said in March 2025: "China will fight till the end" if the US "insists on waging a tariff war, trade war or any other kind of war" (Buchwald, 2025). China initially retaliated by announcing 15% tariffs on wheat, corn, chicken, and cotton imports from the US, and a 10% tariff on pork, sorghum, beef, soybeans, aquatic products, vegetables, fruits, and dairy products. Separately, Alfredo Montufar-Helu, head of the China Center for the Conference Board, said that China's retaliatory tariffs followed a "restrained, targeted approach aimed at causing pain to those industries that matter the most to the supporters of the Trump administration" (Ivuskans, 2025). Trump responded strongly and indicated that these latest tariffs, while substantial, are just the start. The US implemented reciprocal tariffs on April 2, 2025, matching other countries' tariffs dollar for dollar. Although the US and China have later agreed to work on extending a deadline for new tariffs on each other, this illustrates the force of Trumpian capitalism.

6 Conclusion

The twenty-first century has witnessed a fascinating interplay between capitalism and socialism, with both systems undergoing major transformations. In the West, the limitations of neoliberal/shareholder capitalism have become increasingly apparent, prompting a shift towards more inclusive models such as stakeholder capitalism (Schwab & Vanham, 2021a). This new paradigm recognizes the importance of balancing the interests of various stakeholders, including employees, customers, and the environment, alongside those of shareholders. However, the rise of surveillance capitalism presents a new set of challenges, particularly in relation to data privacy and ethical considerations. The ability of tech giants to collect and monetize vast amounts of personal data raises concerns about potential abuses of power and the need for stronger regulatory frameworks (Zuboff, 2019). Time will tell, but it seems to be part of the dark side of progressivism.

In the East, the resurgence of state intervention, exemplified by China's model of state capitalism, has blurred the traditional boundaries between capitalism and socialism. What is interesting and quite ironic about the influence of technological innovation, is that freedom is now more readily sacrificed at the expense of authoritarian tendencies, as can be seen in the apparent merge between socialism and capitalism. This hybrid system, combining elements of market dynamism with centralized control, has not only fuelled China's economic ascent, but also sparked debates about the future trajectory of global capitalism.

China's model of state capitalism, a system that blends market-driven policies with strong state intervention, is a unique approach that challenges conventional definitions and prompts a re-evaluation of established economic paradigms. In view of this, one focus area of the chapter was to determine if the distinct economic systems of socialism and capitalism are in fact converging in some way. The chapter also explored the nuances of 'new socialism', which integrates digital technology, sustainability, and decentralized governance into socialist principles. The analysis further considered how these evolving systems impact the achievement of the SDGs, noting that the synergy between capitalism and socialism, particularly in their progressive forms, can contribute to more balanced and inclusive approaches to economic, social, and environmental challenges. Finally, the chapter addressed the potential disruption caused by the resurgence of Trumpian capitalism, characterized by economic nationalism, deregulation, and protectionism.

Several key insights emerged from this exploration. First, it is clear that Western capitalism is in a state of flux, adapting to pressures from technological change, environmental concerns, and socio-economic disparities (United Nations, 2025). The rise of stakeholder capitalism signals a growing recognition of the need for businesses to consider their broader social and environmental responsibilities. However, the simultaneous rise of surveillance capitalism highlights the potential for new forms of exploitation and the importance of ethical data usage and privacy protections as it leans towards authoritarianism.

Second, China's state capitalism presents a unique model that defies easy categorization. By combining market mechanisms with strong state control, China has achieved remarkable economic growth while maintaining a socialist political structure (Wu, 2022). A comparative analysis of state capitalism and socialism reveals that while there are similarities like a significant degree of state intervention in the economy, they differ in their approaches to ownership, market mechanisms, and social welfare. China's model also raises questions about political freedoms, wealth distribution, and the role of the state in the economy. Third, the chapter identified a potential convergence between progressive versions of capitalism and socialism. Stakeholder capitalism and new socialism share a common emphasis on social equity, innovation, environmental sustainability, and inclusive economic growth. Both frameworks challenge the conventional focus on shareholder value and advocate for a more equitable distribution of value.

Western capitalism has evolved from a singular focus on shareholders to a model that considers broader societal impacts (stakeholder capitalism), while simultaneously embracing data-driven profit mechanisms (surveillance capitalism). Central to this multifaceted transformation of Western capitalism—driven by technological advancements, societal concerns, and geopolitical shifts—is a growing recognition of the need for more inclusive and sustainable economic models. The challenge ahead is to align these competing forces in a way that fosters economic growth, protects individual rights, and ensures ethical digital governance. In an era of changing economic priorities, rapid technological advancements, and new societal values, specifically the following key drivers of progressive change can be identified:

- growing awareness of the negative externalities of shareholder capitalism (e.g., climate change, inequality, and unsustainability);
- pressure from consumers, employees, and activists for corporations to adopt more ethical and sustainable practices;
- the rise of ESG investing, which evaluates companies based on their impact on society and the environment;
- the rise of the internet and digital technologies, enabling unprecedented data collection; and
- the shift from industrial capitalism to a data-driven economy, where data is 'digital gold'.

So, are capitalism and socialism merging? The answer is nuanced. While state capitalism is not the new socialism, progressive versions of both systems are indeed showing signs of convergence. This convergence is driven by a shared desire to address the shortcomings of neoliberal capitalism and to create a more just and sustainable economic system. However, significant differences remain, particularly regarding the role of the state, the ownership of enterprises, and the extent of market regulation. The interplay between these evolving models of capitalism and socialism has significant implications for global economic dynamics and social equity. The rise of stakeholder capitalism in the West suggests a growing recognition of the need for businesses to play a more responsible role in society. In the East, the success of state capitalism in driving economic growth raises questions about the optimal balance between state intervention and market forces.

This exploration holds significant implications for understanding the future of the global economy. The rise of stakeholder capitalism, surveillance capitalism, and state capitalism reflects a broader restructuring of economic power and a growing awareness of the interconnectedness of economic, social, and environmental issues. These trends challenge conventional wisdom and demand a re-evaluation of established economic paradigms. Traditional categories of capitalism and socialism may no longer be sufficient to describe the hybrid models that are emerging in the twenty-first century. Practically, the chapter finds that policies that promote stakeholder engagement, protect data privacy, and foster sustainable and inclusive development are of utmost importance.

The need for international cooperation to address global challenges such as climate change, inequality, cyber risks, and economic instability is also of vital importance. There are some challenges, though. The need to balance innovation with ethical considerations, ensure equitable distribution of wealth, and address pressing global issues, such as climate change and inequality requires, careful navigation. And let us not forget about technological innovation's 'Frankenstein Factor'.[6] The resurgence of Trumpian capitalism, with its emphasis on economic nationalism and protectionism, adds another layer of complexity to the evolving global landscape. Being a defender of democracy and Western freedom against authoritarianism, Trumpian capitalism also contain contradictory elements in that it wants to 'force freedom', especially through its global drive of 'Peace Through Strength' and trade wars. Trump's combative style of politics does present a challenge to navigating global cooperation, but it need not be an insurmountable challenge.

Finally, in the twenty-first century, emerging economic ideologies have further shaped the evolving and multifaceted relationship between capitalism and socialism. While the two systems retain distinct characteristics, they are also showing signs of convergence, particularly in their progressive forms. This convergence offers the potential for a more equitable, sustainable, and inclusive global economy. This is the good/beneficial side of progressive thinking. However, realizing this potential will require overcoming significant challenges, including political resistance, economic inequalities, the potential for exploitation, and authoritarian tendencies at the expense of freedom.

Looking ahead, it is clear that the future of the global economy will depend on our ability to harness the strengths of both capitalism and socialism while mitigating their weaknesses. This will require innovative policies, strong institutions, and

[6] The Frankenstein Factor in technology refers to the unintended and potentially harmful consequences of innovation. The risks that arise when advancements outpace our ability to control or fully understand their implications.

a shared commitment to building a more just and sustainable world. Whether the progressive synergy between capitalism and socialism can withstand the disruptive forces of economic nationalism and political polarization remains to be seen. No one can deny that the new Trump administration is shifting the balance of power—at least to some degree—back to the West. Yet, the trends identified in this chapter suggest that the quest for a more balanced and inclusive economic system will continue to shape the global landscape in the years to come. By fostering a deeper understanding of these evolving trends and models, we can work towards creating a more prosperous, equitable, and sustainable future for all.

References

Alami, I., Dixon, A. D., & Mawdsley, E. (2021). State capitalism and the new global development regime. *Antipode, 53*(5), 1294–1318.

Alami, I. and Dixon, A.D. (2023). Uneven and combined state capitalism. *EPA: Economy and Space*, 55(1), pp. 72–99.

Allen, F., Carlette, E., & Marquez, R. (2009). Stakeholder capitalism, corporate governance and firm value. *EUI Working Paper ECO 2009/10*, pp. 1–43.

Alshamy, Y., Coyne, C. J., Hall, A. R., & Owens, M. (2024). Surveillance capitalism and the surveillance state: A comparative institutional analysis. *Constitutional Political Economy*, 1–23. https://doi.org/10.1007/s10602-024-09438-z

Alexander, F., Ensign-Barstow, H., Palladino, L., & Kassoy, A. (2020). *From shareholder primacy to stakeholder capitalism: A policy agenda for systems change.* B Lab & The Shareholder Commons.

Baer, H. A., & Singer, M. (2023). Planetary health: Capitalism, ecology and eco-socialism. *Capitalism Nature Socialism, 34*(4), 20–38. https://doi.org/10.1080/10455752.2023.2192953

Beck, D., & Ferasso, M. (2023). How can stakeholder capitalism contribute to achieving the Sustainable Development Goals? A Cross-network literature analysis. *Ecological Economics, 204*, 1–14.

Bettelheim, C. (1985). Marxist thinking about the experience of history (La pensée marxienne à l'épreuve de l'histoire, interview in Les Temps modernes), *472*, 20–60.

Billingsley, K. L. (2022). *Klaus Schwab proclaims China a "role model" for many nations.* https://blog.independent.org/2022/12/01/klaus-schwab-china/

Bloom, P. (2023). *Authoritarian capitalism in the age of globalization.* Edward Elgar Publishing.

Boer, R. (2021). *Socialism with Chinese characteristics.* Springer.

Bolton, M. (2020). 'Democratic socialism' and the concept of (post) capitalism. *The Political Quarterly, 91*(2), 334–342.

Boustanifar, H. (2023). An alternative to shareholder capitalism? A review of Alex Edmans' "Grow the pie: How great companies deliver both purpose and profit." *Journal of International Financial Management and Accounting, 34*, 378–383.

Bremmer, I. (2009). State capitalism comes of age: The end of the free market? *Foreign Affairs, 88*(3), 40–55.

Brundenius, C. (2020). *Reflections on socialism in the 21st century.* Springer International Publishing.

Buchwald, E. (2025). *China and Canada immediately retaliate against Trump's tariffs. Mexico is next.* https://edition.cnn.com/2025/03/04/economy/trade-mexico-canada-china-tariffs-trump-hnk-intl/index.html

Bukharin, N. (1925) Toward a theory of the imperialist state (K teorii imperialisticheskogo gosudarstva). *Revolyutsiya prava*, Moskow.

Coalition for Inclusive Capitalism. (2025). *What is inclusive capitalism?* https://coalitionforinclusivecapitalism.com/what-is-inclusive-capitalism/

Cockshott, W. P., & Cottrell, A. (2000). *Towards a new socialism*. Spokesman.

De Jong, M. (2021). Inclusive capitalism: The emergence of a new purpose paradigm in economics and business administration and its implications for public policy. *Global Public Policy and Governance, 1*, 159–174.

Freeman, R. E., Martin, K., & Parmar, B. (2007). Stakeholder capitalism. *Journal of Business Ethics, 74*, 303–314.

Fuchs, C. (2017). Donald Trump: A critical theory-perspective on authoritarian capitalism. *tripleC: Communication, Capitalism & Critique. Open Access Journal for a Global Sustainable Information Society, 15*(1), 1–72.

Gonzalez-Vicente, R., & Cheng, H. (2025). China, geoeconomics and the 'new' state capitalism. *EPA: Economy and Space, 57*(1), 128–132.

Gu, J., Zhang, C., Vaz, A., & Mukwereza, L. (2016). Chinese state capitalism? Rethinking the role of the state and business in Chinese development cooperation in Africa. *World Development, 81*, 24–34.

Guttmann, R. (2023). Can we reform capitalism for its own good? A roadmap to sustainability. *PSL Quarterly Review, 76*(304), 67–88.

Huang, P. H. (2022). Realizing diversity, sustainability, and stakeholder capitalism. *Emory Corporate Governance & Accountability Review, 28*, 28–55.

Hornung, S., Unterrainer, C., & Höge, T. (2022). Dialectics of sustainability: Contrasting mainstream neoliberal and critical ecosocialist perspectives on sustainable development. *European Journal of Sustainable Development Research, 6*(1), 1–9.

Howard, M. C., & King, J. E. (2001). 'State capitalism' in the Soviet Union. *History of Economics Review, 34*(1), 110–126. https://doi.org/10.1080/10370196.2001.11733360

Ivuskans, G. (2025). *Canada, China and Mexico retaliate as Trump's 25% tariffs on goods take effect*. https://www.kpua.net/2025/03/05/canada-china-and-mexico-retaliate-as-trumps-25-tariffs-on-goods-take-effect/

Jaffee, D. (2019). The current crisis of US neoliberal capitalism and prospects for a New "Social Structure of Accumulation." *Review of Radical Political Economics, 51*(2), 193–210.

Kautsky, K. (2019). *Democracy and republicanism*. Brill.

Kavenna, J. (2019). *Shoshana Zuboff: 'Surveillance capitalism is an assault on human autonomy'*. https://www.theguardian.com/books/2019/oct/04/shoshana-zuboff-surveillance-capitalism-assault-human-automomy-digital-privacy

Khalili, S. (2023). *It's Presidents Day, so we asked: Is the president of the United States still the most powerful person on earth?* https://news.northeastern.edu/2023/02/16/presidents-day-2023/

Kim, R. C. (2022). Rethinking corporate social responsibility under contemporary capitalism: Five ways to reinvent CSR. *Business Ethics, Environment & Responsibility, 31*, 346–362.

Lane, D. (2009). Global capitalism and the transformation of state socialism. *Studies in Comparative International Development, 44*, 97–117.

Lane, D. (2013). *The capitalist transformation of state socialism: The making and breaking of state socialist society, and what followed*. Routledge.

Lebowitz, M. A. (2013). The state and future of socialism. *Socialist Register, 49*, 345–368.

Lebowitz, M. A. (2016). What is socialism for the 21st century? *Monthly Review, 68*(5), 26–36.

Lenin, V. (1917). Imperialism, the highest stage of capitalism. *Petrograd*, 1–99

Lin, Y., Fu, X., & Fu, X. (2021). Varieties in state capitalism and corporate innovation: Evidence from an emerging economy. *Journal of Corporate Finance, 67*, 1–41.

Lin, L. Y., & Milhaupt, D. J. (2023). China's corporate social credit system: The dawn of surveillance state capitalism? *The China Quarterly, 256*, 835–853. https://doi.org/10.1017/S030574102300067X

Mazzucato, M. (2024). Collective value creation: A new approach to stakeholder value. *International Review of Applied Economics, 38*(1–2), 43–57. https://doi.org/10.1080/02692171.2022.2144149

McGaughey, E. (2018). Fascism-lite in America (or the Social Ideal of Donald Trump). *British Journal of American Legal Studies, 7*(2), 291–315.
Mhlanga, D. (2022). Stakeholder capitalism, the Fourth Industrial Revolution (4IR), and sustainable development: Issues to be resolved. *Sustainability, 14*(7), 1–20.
Michaels, J. D. (2020). We the shareholders: Government market participation in the postliberal U.S. political economy. *Columbia Law Review, 120*(20), 465–546.
Millán-Acevedo, N., & Gómez-Bruna, D. (2025). Contradictions between capitalism and sustainable human development: Where is the 2030 agenda headed? *Capitalism Nature Socialism*, 1–20.
Mollan, S., & Geesin, B. (2020). Donald Trump and Trumpism: Leadership, ideology and narrative of the business executive turned politician. *Organization, 27*(3), 405–418.
Mosco, V. (2014). *To the cloud: Big data in a turbulent world*. Routledge.
Musacchio, A., Lazzarini, S. G., & Aguilera, R. V. (2015). New varieties of state capitalism: Strategic and governance implications. *Academy of Management Perspectives, 29*(1), 115–131.
National Geographic. (2025). *Socialism*. https://education.nationalgeographic.org/resource/socialism/
Nimtz, A. H. (2019). *What is unmistakable about Trump: His naked capitalism*. https://www.minnpost.com/community-voices/2019/09/what-is-unmistakable-about-trump-his-naked-capitalism/
Norman, G. (2025). *Trump responds to Trudeau calling tariffs 'very dumb' as trade war ramps up*. https://www.foxbusiness.com/economy/trudeau-tells-trump-we-not-back-down-from-trade-war-calling-tariffs-very-dumb
Pierleoni, M. R. (2020). Capitalism after the coronavirus: Public goods, sustainability and the role of the state. *Theoretical Economics Letters, 10*(4), 868–874.
Pollock, F. (1936). *State capitalism: Its possibilities and limitations*. Routledge.
Poulantzas, N. A., & Hall, S. (1978). *State, power, socialism*. Nlb.
Samans, R., & Nelson, J. (2022). *Sustainable enterprise value creation: Implementing stakeholder capitalism through full ESG integration*. Palgrave Macmillan.
Scholz, T. (2016) *Uberworked and underpaid: How workers are disrupting the digital economy*. Wiley.
Schwab, K., & Vanham, P. (2021a). *Stakeholder capitalism: A global economy that works for progress, people and planet*. Wiley.
Schwab, K., & Vanham, P. (2021b). What is the difference between stakeholder capitalism, shareholder capitalism and state capitalism? *WEF*. https://www.weforum.org/stories/2021/01/what-is-the-difference-between-stakeholder-capitalism-shareholder-capitalism-and-state-capitalism-davos-agenda-2021/
Silverman, B. C., & Ingram, P. (2017). Asset ownership and incentives in early shareholder capitalism: Liverpool sipping in the eighteenth century. *Strategic Management Journal, 38*, 854–875.
Sperber, N. (2019). The many lives of state capitalism: From classical Marxism to free-market advocacy. *History of the Human Sciences, 32*(3), 100–124.
Sutterlütti, S., & Meretz, S. (2023). *Make capitalism history: A practical framework for utopia and the transformation of society*. Springer.
Sweezy, P. (1946). *The theory of capitalist development*. D. Dobson.
Sweidan, O. D. (2021). State capitalism and energy democracy. *Geoforum, 125*, 181–184.
United Nations. (2025). *World economic situation and prospects: February 2025 Briefing, No. 187*. https://www.un.org/development/desa/dpad/publication/world-economic-situation-and-prospects-february-2025-briefing-no-187/
Ullah, S., Adams, K., Adams, D., & Attah-Boakye, R. (2021). Multinational corporations and human rights violations in emerging economies: Does commitment to social and environmental responsibility matter? *Journal of Environmental Management, 280*, 1–13.
Van Niekerk, A. J. (2022). *The inclusive economy: Criteria, principles and Ubuntu*. UJ Press.
Vu, A. N., & Long, G. (2023). Universalism and national ownership in the context of the Sustainable Development Goals (SDGs): perspectives from Vietnam. *International Development Planning Review, 45*(1), 41–66.

Wright, M., Wood, G., Musacchio, A., Okhmatovskiy, I., Grosman, A., & Doh, J. P. (2021). State capitalism in international context: Varieties and variations. *Journal of World Business, 56*(2), 1–16.

Wu, G. (2022). *China's common prosperity program: Causes, challenges, and implications*. The Asia Society Policy Institute.

Xing, L., & Shaw, T. M. (2013). The political economy of Chinese state capitalism. *Journal of China and International Relations, 1*(1), 1–26.

Zanoni, P., & Pitts, F. H. (2022). Inclusion through the platform economy? The 'diverse' crowd as relative surplus populations and the pauperisation of labour. In E. Ness (Ed.), *The Routledge handbook of the gig economy* (pp. 33–46). Routledge.

Zuboff, S. (2019). *The age of surveillance capitalism: The fight for a human future at the new frontier of power*. PublicAffairs.

The New Global Economy and Economic Inclusion: Achieving the SDGs Together

5

Abstract

This chapter provides a comprehensive analysis of the new global economy of the twenty-first century, characterized by unprecedented technological advancements, shifting geopolitical power, and the evolving dynamics of globalization. It explores the intersection of economic inclusion and sustainable development, emphasizing their role in achieving the UN SDGs by 2030. The chapter first examines the 4IR and its transformative impact on industries, labour markets, and economic productivity. It highlights how digitalization, automation, and AI are restructuring global value chains, requiring new policies to ensure inclusive growth. It also examines other key trends like the geopolitical and geoeconomic shifts, particularly the US-China tech and trade war, de-dollarization, and regional economic fragmentation. The chapter also explores the role of financial innovation, including FinTech, blockchain, and green finance, in reshaping global markets and fostering economic sustainability. The growing influence of the BRICS+ nations, the rise of economic nationalism, and the emergence of re-globalization are evaluated as key dynamics shaping the future global order. Ultimately, the chapter finds that the current global economic landscape necessitates a move away from purely growth-driven models towards inclusive and sustainable economies that prioritize the well-being of people and the planet. It also finds that achieving the SDGs requires strong political will, increased financing, digital transformation, local participation, and global partnerships. It calls for a cooperative, innovation-driven approach to ensure that economic progress benefits all, particularly in emerging markets. The chapter concludes by advocating for a shared vision between the East and West centered on inclusive growth, sustainable development, and cooperation as the path forward for a resilient and equitable global economy. The findings reinforce the need for strategic global governance and policy adaptation to navigate the complexities of the modern economy. The chapter's significance lies in its call for a fundamental reorientation of economic thinking and practice to address the intertwined crises of inequality and environmental degradation, aligning the

A. J. van Niekerk, *West to East: A New Global Economy in the Making?*,
https://doi.org/10.1007/978-3-031-93267-0_5

global economy with the SDGs for a more prosperous and sustainable future for all.

Keywords

4IR • Globalization • Financial innovation • Economic sustainability • SDGs • Inclusion

1 Introduction

In the wake of the COVID-19 pandemic and the subsequent energy and food crises, exacerbated by the Russian invasion of Ukraine and instability in the Middle East, the world continue to face economic shockwaves of existential proportions. Even prior to these events, the global economy was showing signs of strain, struggling to maintain sustainable progress and equitable economic systems. Exclusion and inequality have become critical vulnerabilities in the prevailing neoliberal economic framework. This necessitates a shift towards more inclusive and balanced economic models. The rise of the 4IR, coupled with increasing concerns about economic sustainability and financial stability, present unprecedented opportunities and complex challenges. New geopolitical tensions in the West (the US and NATO) are rebalancing (almost reversing) the geoeconomic trends of the past decade, especially in view of China's rapid rise. While global trade, digitalization, and financial innovation have fuelled economic growth, they have also deepened disparities between nations and within societies. Against this background, the question arises: How can the global economy transition toward a more inclusive and sustainable model that benefits all?

In answering this question, the chapter explores the intersections of economic inclusion, financial innovation, and geopolitical shifts as key trends in shaping the future of the global economy. It examines the role of technological disruption, economic nationalism, and sustainability-driven policies in redefining economic models. A particular focus is placed on the role of BRICS+, de-dollarization, digital transformation, and green finance in challenging traditional Western-dominated financial structures and fostering alternative economic paradigms. Unlike previous eras, this new global economy is largely shaped by the 4IR, characterized by the fusion of digital, biological, and physical technologies such as AI, automation, and blockchain. Coupled with a shifting geopolitical landscape marked by the tech and trade war between the US and China, the rise of economic nationalism, and the urgent need for economic sustainability and alignment with the UN's SDGs, the global economy faces a period of rapid change and uncertainty. This chapter aims to provide a comprehensive analysis of these transformative forces and their intricate interplay, exploring their profound implications for achieving economic inclusion and the SDGs.

To navigate this complex landscape, the chapter draws on heterodox economic theories, including institutional, evolutionary, and ecological economics, seeking to provide a holistic understanding of the forces shaping the new global economy. It also draws upon a conceptual framework that integrates the understanding of the 4IR as a driver of both opportunity and disruption, the principles of inclusive economy, which prioritize equitable access to economic opportunities and resources, and the Agenda 2030 for Sustainable Development, providing a roadmap for global sustainability. This framework allows for an examination of how technological advancements and geopolitical shifts are reshaping economic structures and globalization dynamics, and how these changes can either hinder or facilitate the progress towards a more inclusive and sustainable global economy.

The significance of this chapter lies in its timely analysis of the converging forces shaping the new global economy and their crucial implications for achieving a more equitable and sustainable future. By offering a comprehensive overview and integrating various interconnected trends, the chapter contributes to a deeper understanding of the challenges and opportunities in aligning global economic development with the goals of economic inclusion and the SDGs. It highlights the risks posed by deepening financial divides, technological disruption and economic fragmentation, while offering strategies to harness innovation, policy adaptation, and cross-border collaboration for long-term resilience. The chapter contributes to ongoing discussions on redefining global economic governance in an era of uncertainty and transition. It argues that the future of the global economy depends on a shift toward inclusive economic policies, sustainable financial practices, and cooperative global governance (Moch, 2024). Rather than retreating into protectionism and financial exclusivity, economies must embrace new paradigms of shared prosperity, technological inclusivity, and environmental responsibility. Only through collective action and forward-thinking strategies can the global economy achieve sustained progress and stability in the decades ahead.

By analysing technological advancements, geopolitical tensions, and financial shifts, the chapter provides a holistic understanding of how economic inclusion and sustainability can shape the future global order. The findings offer policy recommendations for governments, businesses, and international organizations seeking to balance innovation, equity, and environmental responsibility in economic policymaking. It emphasizes the urgent need for a fundamental reorientation of economic thinking and practice, advocating for a shared global vision rooted in cooperation and the prioritization of human and planetary well-being over narrow economic growth metrics. Ultimately, it seeks to inform policymakers, business leaders, and researchers about the transformative pathways required to navigate the complexities of the twenty-first century global economy and build a more prosperous and sustainable world for all. As the world navigates an increasingly complex and fragmented economic landscape, this chapter calls for a new economic paradigm—one that is collaborative, inclusive, and resilient. By fostering global cooperation, ethical business practices, and strategic governance, the vision of an inclusive global economy can be realized.

2 The New Global Economy of the Twenty-First Century

The innovative knowledge-driven global economy of the twenty-first century is unique due to a combination of unprecedented technological advancements, shifting geopolitical power, and evolving economic structures and paradigms. This economy is characterized by rapid change, uncertainty, and opportunities for innovation and collaboration. Unlike previous eras, it is driven by the Fourth Industrial Revolution (4IR), where AI, automation and blockchain are transforming industries. The tech war between the US and China, alongside trade conflicts and economic nationalism, is reshaping global supply chains and financial systems. The rise of digital currencies is challenging traditional financial models, while the potential for de-dollarization signals a shift in global economic power. In addition, globalization is undergoing transformation, with trends of de-globalization, regionalization, and re-globalization emerging in response to economic and political pressures. One area of growing concern is the need for economic sustainability and alignment with the UN's SDGs, influencing global policies, and pushing for greener and more inclusive economic models. These factors are just a glimpse of what make the twenty-first century economy highly dynamic, interconnected, and complex. The sections that follow will delve deeper into the various aspects.

2.1 4IR and the Technology Revolution

The new global economy is built on a culture of innovation. New ideas, new technologies, and new initiatives represent the signature mark of the twenty-first century. The 4IR, characterized by the fusion of digital, biological, and physical technologies, is transforming industries and societies. The rise of digital technologies has reduced the economic costs of data search, storage, computation, and transmission, which has enabled new economic activities (Acs et al., 2021). The role of information and communications technology in the new 4IR economy has been pivotal, particularly the changing structure of international production. Firms are integrating the production and marketing of goods and services across national borders. The 4IR is driving increased automation, data-driven decision-making, and the creation of new business models. It involves the digitalization of the modern world and the inclusion of digital industry in major industrial sectors. Through Industry 4.0, the 4IR is introducing disruptive technologies like AI, robotics, IoT, blockchain, big data, cloud computing, biotechnology and 3D printing—all of which are changing social, economic and political systems around the world (Nyagadza et al., 2022).

4IR dynamics are prompting changes to emerging economies' industrial innovation. In the 4IR, digital global value chains are optimized with interconnected network systems, which affect how industries in emerging economies operate. Digital technologies and intelligent systems have enabled firms to process more information at drastically reduced costs and given them greater ability to manage complex organizational structures, including extended and dispersed production and value chains (Pandey et al., 2023). Digital transformation encourages adaptable, responsive, and connected businesses that can make more informed choices. The rise of global production networks has been a distinctive characteristic of the digital revolution. Half of all production happens within global innovation networks, where the digitalization of the vertical production system complements the horizontal networking across the entire value chain (Knell, 2021). The development of 4IR technologies is expected to create new opportunities for emerging economies' growth and development, increasing the quality of life and income per capita. Industry 4.0 allows enterprises to enhance their operations by employing intelligent technology-supported machinery. As a result, new economic activities and opportunities emerge from connecting individuals, businesses, devices, data, and operations through digital technology. Relying on digital technology and online transactions, the digital economy is transforming traditional industries. Data is becoming a critical economic asset (digital gold) that drives innovation and competition.

In this knowledge-based economy, skills, qualifications, and experience influence a country's competitive advantage in the global arena. An educated society possesses more specialized knowledge that allows for the development of new technologies, products, and services, which lays the foundation for trade specialisation and attracting foreign investments in local industries (Kisiel & Lubowiedzka, 2010). A country's human resources and the economic value of its human capital endowment, reflected in the educational attainment and technical skills of its population, is essential for empowering the new economy and facilitating the integration of labour in knowledge-based industries (Passaris, 2006). In the 4IR, we are witnessing the rise of 'neo-industrialization', which is an innovative economy based on knowledge, human resources, and networks—dominated by computers, communications, and network technology (Matyushok et al., 2021). Digitalization in the era of the 4IR has a wide range of effects on the job market, including transformations in job roles, skills required, and the overall structure of employment. More specifically, these include:

- *Job creation and destruction*: New jobs are being created and old ones are destroyed. This implies a decline in demand for manual workers and an increased demand for information/knowledge workers. Automation leads to increased productivity, which puts downward pressure on product prices and increases real wages and purchasing power. It also increases the demand for labour that complements these technologies—unfortunately, at the expense of those workers who do not have higher technical skills. Digital technologies enable new employment possibilities, which includes responsibilities in big data analysis, data science, and digital marketing.
- *Changing skill requirements*: The IT revolution has increased the demand for non-routine skills at the expense of jobs demanding routine skills (Hulten, 2015). Digital technologies require the training of digitally-orientated citizens.

New technologies could also lead to upskilling and reskilling if machines substitute unskilled labour and require more skilled labour to operate them. Moreover, job polarization might occur, with simultaneous growth of high-education and low-education jobs and a fall in employment in the middle of the skills distribution. The 4IR is creating entirely new job categories that did not exist before, such as AI ethicists, data scientists, and IoT specialists.
- *A change in labour composition*: The automation of certain work tasks and the shifts in consumer demand affect the demand for labour in specific occupations. Digital technologies (robotics, AI and machine learning) can now complete standardized tasks with enormous efficiency, leading to a displacement of routine manual work and routine office functions by automated electronic systems (e.g., manufacturing and data entry) (Scott, 2011). This reduces the demand for low-skilled labour. The rise of digital platforms has facilitated the growth of the gig economy,[1] where short-term contracts or freelance work opportunities are common. This is changing the traditional employer-employee relationship.
- *Impact on different sectors and work environments*: The digital revolution affects labour markets through a great variety of technologies (Deschacht, 2021). For instance, the number of secretaries has been declining rapidly as executives started using personal computers and smartphones. In manufacturing, some blue-collar occupations are automated by the use of industrial robots in a variety of routine tasks. So also is traditional retailing in decline because of the rise of e-commerce. In the work environment, advances in communication technologies have made remote work more feasible, changing where and how people work. The use of collaborative tools and platforms is becoming more prevalent, requiring workers to be proficient in digital communication and project management tools. The 4IR is making the labour market more global. Companies can tap into talent pools from around the world, increasing competition but also opportunities for workers. Automation and digitalization is also changing the dynamics of outsourcing and offshoring, as some tasks that were previously offshored may now be automated locally.

The 4IR is fundamentally altering the composition of labour by driving demand for high-skilled workers, creating new job categories, and necessitating lifelong learning and adaptation. While it offers opportunities for innovation and efficiency, it also poses challenges related to job displacement, income inequality, and the need for comprehensive policy responses to ensure inclusive growth. The world economy is being restructured by technological changes and the geographic movement of all factors of production. This mobility is already changing the location of production as well as the direction and volume of flows of trade and investment among cities.

[1] The gig economy offers both workers and employers greater flexibility. Instead of traditional full-time employment, workers in the gig economy take on 'gigs' or individual tasks, projects, or jobs, often on a temporary or contract basis.

Potential negative effects of the 4IR are expected to be more impactful in emerging economies given the disadvantages it creates in global value chains. Emerging economies may face challenges as digitalisation may cause serious job losses in all industries (Nyagadza et al., 2022). It is vital that the challenges and concerns about the digital revolution, especially in view of surveillance capitalism, China's mixed-socialism, and the increased risk of cyber threats and attacks, must be highlighted to proactively develop countermeasures to mitigate possible negative consequences. The sheer volume of data collected through every facet of our digital interactions raises ethical concerns and questions around data privacy.

Another concern is that the digital economy or web economy is too dependent on internet access (Javaid et al., 2024). Particularly businesses and services that rely entirely on the internet are vulnerable to disruptions caused by technical failures, cyberattacks, or natural disasters. While the digital economy offers immense opportunities for growth, innovation, and connectivity, the world is arguably becoming too reliant on digital infrastructure. Internet outages can halt operations, leading to significant financial losses and reduced productivity. Overdependence on the internet creates a single point of failure, increasing systemic risk and making economies less resilient to shocks. Without robust offline alternatives, disruptions in internet access can have cascading effects across multiple sectors. Addressing these concerns requires investment in infrastructure, policies to bridge the digital divide, enhanced cybersecurity measures, and strategies to ensure economic resilience and inclusivity.

2.2 Geopolitical and Geoeconomic Shifts

The tech war between the US and China began as a trade war between them in 2018, halting hyper-globalization and amplifying divisions in the world economy. The US-China trade conflict started with an investigation into China's acts related to technology transfer, intellectual property, and innovation (Tung et al., 2023). The trade conflict escalated when the US prohibited the use of information and communications technology and services from 'foreign adversaries' due to national security threats. As a result, economic statecraft were increasingly used through trade restrictions to protect intellectual property and address forced technology transfer (Aggarwal & Reddie, 2021). The US and China use national security rationales for industrial policy, especially in dual-use technologies like AI, quantum computing, and cybersecurity. Both countries have imposed barriers on the import and export activities of each other's companies and restrictions on cross-border investments.

Such geopolitical tensions have polarized the world, creating increased strain between East and West. Particularly the competitive tension between the US and China are forcing countries, businesses, and institutions to navigate an environment of strategic rivalry in different areas. The US-China trade war has led to protectionist policies, tariffs, and reshoring efforts that disrupt global supply chains. Nations are pressured to align with either China's BRI or US-led economic frameworks.

The battle over AI, semiconductors, and 5G technology has created technological decoupling as the US and China are creating competing technological ecosystems (Buckley, 2023). On the military front, the rise of security alliances such as AUKUS (Australia-UK-US) and the QUAD (the Quadrilateral Security Dialogue between the US, Australia, India, and Japan) counterbalances China's growing influence in the Indo-Pacific, while China is strengthening military ties with Russia and expands its presence in the South China Sea. In terms of global governance and influence, institutions like the UN, WTO, and IMF are sites of competition for influence. The Global South faces growing pressure to choose between Western-led governance models and China's state-driven development approach. Ideological and political polarization is another area of strategic rivalry. Democracies and authoritarian regimes are increasingly positioned as opposing models. The US promotes a rules-based international order, while China advocates for non-interference and state sovereignty (Staiger, 2022). As the world is shifting toward a fragmented, multipolar order, where alliances are being redrawn, and economic, technological, and political divides deepen. Many countries are attempting to maintain strategic autonomy, but growing US-China tensions make neutrality increasingly difficult.

The result of escalating geopolitical tension is geoeconomic fragmentation—that is, the growing division of the global economy into competing blocs. Apart from the ways already mentioned, this can also be seen in a number of other ways. For example, the Ukraine war has led to Europe diversifying away from Russian energy, while Russia strengthens trade with China, India, and the Global South. Another way is the 'splinternet': the internet, which underpins global commerce, is itself fracturing (Buckley, 2023). Different 'internets' are co-existing with distinct ethical visions, technological realities, and institutional support. This splintering affects the free flow of data and disrupts global value chains. There is also a trend away from globally integrated or matrix organizations toward multi-domestic operations that emphasize autonomy in each territory, with looser governance structures across fragmented territories (Chen et al., 2023). Economic policies designed to fracture the world economy have costs for the target, the instigator, and the global economy as a whole. The WTO, in particular, faces serious challenges, including the failure of the Doha Round and a crisis in its appellate body, limiting its ability to manage economic statecraft (Aggarwal & Reddie, 2021). Conflicting laws, regulations, and jurisdictions arising from geoeconomic fractures can lead to a crisis of legitimacy in international commerce. In terms of AI and data protectionism, countries are enforcing stricter data localization laws, causing further fragmentation in the global digital economy.

Bosone et al. (2024) did an excellent study by presenting new evidence and quantifying the timing and impact of geopolitical tensions as shaped by trade flows from 2000 to 2022. To do so, they used the latest developments in trade gravity models. Geopolitics is proxied by the geopolitical distance between country pairs. They illustrate the historical trajectory of geopolitical distance across four country pairs: the US-China, US-France, Germany-China, and Germany-France. They reveal that both the US and Germany have maintained a consistently greater geopolitical distance from China, with this gap widening further in recent years.

Geopolitical distance is incorporated into a standard gravity model with a comprehensive set of fixed effects, allowing for the control of unobservable factors influencing trade. The analysis also accounts for international border effects and bilateral, time-varying trade cost variables, such as tariffs and trade agreements. To mitigate potential bias from the use of energy flows as political leverage by rival nations, the study focused on manufacturing goods excluding energy as the dependent variable. The findings are based on three-year average data. They found that geopolitics began to significantly influence global trade only after 2018, aligning with the onset of the US-China tariff war and later exacerbated by Russia's invasion of Ukraine. The analysis highlights the varying impact of geopolitical distance across different country groups, providing strong evidence of friend-shoring, while no clear indications of near-shoring[2] were observed. Lastly, the study demonstrates that geopolitical factors are increasingly shaping EU trade, particularly concerning strategic goods.

On the one hand, the Global South is leveraging these geopolitical tensions to gain influence and negotiate better trade and investment terms. On the other hand, they and others have no choice but to resort to specific strategies to navigate the fragmentation. Companies are introducing 'end-to-end' transparency in value chains to minimize exposure to shocks and enable rapid responses (Buckley, 2023). Corporations are increasingly using scenario planning to anticipate and mitigate the impacts of potential shocks. As a form of corporate diplomacy, companies are emphasizing identification with the countries in which they operate, requiring a downplaying of ownership identity and a positive alignment with local contexts. These adjustments are just a few examples of how firms are dealing with the new geopolitical reality. Clearly, geoeconomic fragmentation is causing a shift away from a seamlessly integrated global economy towards a more regionalized and contested landscape, influenced by national policies, security concerns, and the strategic realignments of economic powers.

Countries also use soft power, economic diplomacy, and cultural influence as economic tools to strengthen their geopolitical positions by shaping global perceptions, building strategic alliances, and expanding their economic influence. These strategies aim to build relationships and promote specific economic and political agendas (Aggarwal & Reddie, 2021). Soft power refers to a nation's ability to influence others through cultural appeal, values, and diplomacy rather than coercion or military force. As part of 'cultural exports' countries like South Korea leverage K-pop, K-dramas, and tech brands (Samsung and Hyundai) to amplify national identity and attract investment. The US and the UK use prestigious universities like Harvard and Oxford to attract global talent and strengthen international ties. English as a global business language strengthens the US and UK's economic

[2] Near-shoring is the practice of relocating business operations, manufacturing, or supply chains to a nearby country, typically one that shares a border or is within the same region (Capello & Dellisanti, 2024). For example, US companies may near-shore production to Mexico instead of outsourcing to distant countries like China, benefiting from lower labour costs while maintaining geographical proximity and reducing supply chain risks.

dominance. Countries like Qatar (2022 FIFA World Cup) and China (Beijing Olympics) use global events to boost economic standing and international prestige.

In terms of global media and narratives, China's CGTN, Russia's RT, and the US's Hollywood shape global opinions and promote national interests. Countries construct narratives to portray themselves favourably. For instance, China presents itself as a benevolent regional power to neighbouring Asian countries, emphasizing mutual benefits and long-term relationships. To the West, China promotes 'win–win' scenarios and 'special relationships' (Buckley, 2023). In terms of economic diplomacy, China, for example, finances infrastructure projects in developing nations as part of its BRI, creating economic dependencies and securing access to natural resources. Japan uses investment and strategic negotiations to maintain its competitive edge against China and Russia. The US and EU provide development aid, often with political and economic conditions attached. The EU's free trade agreements and the US-Mexico-Canada Agreement (USMCA) build economic networks that align with geopolitical goals. Japan and Germany export high-tech industries and innovation as a soft power tool. The US relies on its global brands, media dominance, and military alliances, while China uses various types of investments, cultural diplomacy (e.g., Confucius Institutes), and trade to expand influence. The EU leverages trade deals, regulations, and green technology leadership to shape global policies. Russia uses its oil and gas exports as leverage over Europe and other energy-dependent regions. By strategically integrating soft power, economic diplomacy, and cultural influence, countries can shape global trade, alliances, and geopolitical dynamics in their favour.

2.3 Globalization Dynamics and Systemic Risk

Since the global financial crisis, trade has been expanding at a slower pace than GDP, marking the onset of an era often referred to as 'slowbalisation' (Antràs, 2020). While this deceleration could be seen as a natural progression following a period of rapid trade growth, it has been accompanied by a sharp rise in trade restrictions. The US-China tariff war intensified this trend, and geopolitical tensions, particularly following Russia's invasion of Ukraine, have further fuelled discussions on protectionism, near-shoring, and friend-shoring as strategies to safeguard economic interests (Goldberg & Reed, 2023). The rise of China and its geostrategic competition with the US have weakened the trend of globalization. The clash of interests between these nations has broad and long-term implications for the global economy. Geoeconomic fragmentation is fundamentally altering globalization, creating a more divided world economy characterized by regional alliances, financial fragmentation, and technological bifurcation (Bolhuis et al., 2023).

Taking on different forms, globalization has indeed reshaped the global economy of the twenty-first century, driving interconnectedness, economic shifts, and geopolitical realignments, as well as significant transformations across various sectors. It has served as a catalyst for economic growth, technological advancements, altered business practices, and intensified interdependence among nations. Globalization in the twenty-first century is characterized by the increasing integration of a variety of small and medium-sized players in international networks of production, movement of goods, and flows of information and knowledge (Mathews, 2006). Here are key aspects of globalization's role in shaping the new global economy:

- *Increased interdependence and integration*: Globalization has fostered growing interdependence among countries through increased cross-border transactions in goods, services, and capital (Regnerova et al., 2021). The integration of national economies into a vast, interconnected global economic system has led to a surge in global trade and business. Globalization has led to the creation of complex global supply chains, where production processes are spread across multiple countries to optimize costs and efficiency. Countries are now more reliant on each other for trade, investment, and resources. Economic events in one country (e.g., recessions and pandemics) can have ripple effects globally, so increasing systemic risk. FDI has increased as MNCs invest in foreign markets, leading to the transfer of capital, technology, and expertise across borders.
- *The rise of MNCs and the transformation of business and corporate strategies*: As part of their global expansion, MNCs operate in multiple countries, leveraging globalization to access new markets, cheaper labour, and resources. MNCs often wield significant influence over global trade, investment, and even political decisions, shaping the global economic landscape (Tung et al., 2023). Globalization has intensified competitive rivalry and reduced barriers to entry in markets across the world. Big tech corporations, in particular, wield unprecedented economic and political influence through data ownership and AI-driven decision-making. MNCs have adopted new strategies like accelerated internationalization and strategic innovation to succeed in the global economy. The rise of "dragon multinationals" from the Asia Pacific region exemplifies the increasing integration of small and medium-sized players in international networks (Mathews, 2006:6).
- *New economic models*: New approaches to the economy, such as green, circular, and bio-economies, have become priorities in many countries in response to evolving challenges and opportunities. New models now integrate environmental factors, promote sustainable practices, and leverage technological advancements (Trushkina, 2022). More examples include: the platform economy (where businesses operate through digital platforms, leveraging data, AI, and network effects); the knowledge-based economy (emphasizing the importance of innovation and human capital); blockchain economy (enabling decentralized and transparent transactions); and hybrid economies (blending elements of capitalism, socialism, and state intervention). There is a growing emphasis on achieving sustainable development goals and addressing climate change.
- *Financial globalization*: The integration of financial markets allows for the free movement of capital, enabling investment and economic growth but also increasing vulnerability to financial crises (i.e., systemic risk). Globalization

has led to the rise of foreign exchange markets, where currencies are traded around the clock (globally), impacting trade and investment. In addition, the globalization of finance, digital banking, and decentralized finance (DeFi) are challenging traditional banking systems. Due to globalization, nations like Saudi Arabia, Norway, and China use SWFs to invest strategically across the globe.

Globalization has fostered unprecedented levels of economic integration, technological advancement, and shifts in economic power. However, it also faces challenges related to sustainability, fragmentation, inequalities, economic vulnerabilities, and the need for effective global governance. Currently, the debate between open markets vs. economic nationalism is one of the most significant issues reshaping global trade policies in the twenty-first century. This tension reflects competing visions for how economies should operate in an interconnected world. The debate is challenging decades of globalization and influencing how countries engage in global commerce (Bolhuis et al., 2023). This debate has intensified due to geopolitical conflicts, trade wars, supply chain disruptions, and rising populism, leading to shifts in global economic strategies.

While economic nationalism is on the rise (e.g., Trumpian capitalism), full-scale de-globalization remains unlikely. Instead, the future of trade will be more regional, strategic, and politically driven, with technology, energy security, and geopolitics playing a defining role in economic policies. For example, the fragmentation of global supply chains is leading to regionalization and friend-shoring strategies in response to economic nationalism. The evolution of trade agreements is resulting in the prioritization of digital and service-based economies. Although open markets promote growth and interdependence, economic nationalism emphasizes resilience and sovereignty. The challenge for policymakers is to strike a balance that maximizes the benefits of globalization while addressing its downsides to ensure a fair and sustainable global trading system (Staiger, 2022).

The world is undergoing shifts in degrees of openness and integration, reflecting the co-evolution of government initiatives and policies, technological advancements, and the competitive dynamics of MNCs. Periods of globalization have been replaced by periods of friction and fracturing of the global economy, including significant alterations of the strategies and organization of MNCs (Tung et al., 2023). As a result, re-globalization emerged as an alternative to de-globalization. This is where supply chains are being moved to more friendly countries, and trade is being focused on those friendly nations. Re-globalization is driven by digital connectivity, resilient trade networks and AI-powered trade, green finance, and inclusive growth (Goldberg & Reed, 2023). With re-globalization, economic integration continues but in a more regionalized, strategic, and politically driven manner. Unlike the earlier phase of globalization (1990s and early 2000s), which was characterized by hyper-globalization, free trade, and economic liberalization, re-globalization acknowledges the shifts caused by geopolitical tensions, supply chain disruptions, technological advancements, and economic nationalism while still promoting interconnected economies. It is important to note that re-globalization represents an evolution, not a reversal, of globalization. Instead of a return to economic nationalism, re-globalization is about adapting globalization to new realities, balancing efficiency, security, and sustainability.

Re-globalization represents a shift in how countries and businesses approach global integration, emphasizing resilience, sustainability, and inclusivity. Unlike hyper-globalization, which focused heavily on maximizing efficiency and cost reduction through global supply chains, re-globalization seeks to address the vulnerabilities of hyper-globalization and aims to create a more balanced and sustainable global economic system (Nahm, 2021). By focusing on regionalization, digital trade, and green practices, re-globalization aims to build a system that is better equipped to deal with the complexities of the twenty-first century. To stimulate inclusive economic growth, there is a greater focus on ensuring that the benefits of globalization are more evenly distributed, both within and between countries. Countries are designing policies to support developing economies and marginalized communities. Re-globalization emphasizes sustainable practices, such as reducing carbon footprints and promoting circular economies. It also prioritizes fair trade and ethical labour practices, making them integral to global supply chains.

In terms of global trade coordination, the impact of the 4IR on global competitiveness and the structural transformation of global trade and investment patterns will require new regulatory frameworks (Acharya et al., 2019). What is complicating regulations is the fact that different regions and countries have varying standards for trade, sustainability, and digital governance. However, harmonizing these standards is essential for seamless re-globalization. There are significant challenges that re-globalization is facing, such as increased protectionism, shifts in supply chain strategies, challenges to multilateral institutions, and increased complexity for international businesses. The long-term implications of these trends remain uncertain, but the shift away from hyper-globalization towards re-globalization—even though in a more fragmented and contested global economic landscape—presents an opportunity for new economic progress. The interplay between these forces continues to shape the trajectory of the twenty-first century global economy.

2.4 Financial Innovation

Financial innovation in the twenty-first century is reshaping global finance through technology-driven advancements, new financial instruments, and regulatory changes. These innovations aim to improve efficiency, accessibility, and security while also creating new challenges and risks. It is characterized by technology-enabled transformations in the delivery and management of financial services. These advancements have spurred increasing disintermediation in financial services, leading to changes in regulatory and supervisory approaches (De Koker et al., 2019). Being a wave of innovation, there are a number of key areas in which financial innovation is taking place.

Firstly, financial technology (FinTech) is revolutionizing the banking industry by making financial services faster, more efficient, and more accessible. Traditional banks are facing growing competition from FinTech start-ups, which leverage technology to offer digital-first financial solutions (Tarawneh et al., 2024). It integrates software, data analytics, AI, blockchain and cloud computing to enhance banking, payments, lending, insurance, investment, and more. FinTech has emerged as a ground-breaking force, catalysing innovation and propelling economic growth by challenging established financial paradigms, broadening the accessibility of financial services, and stimulating entrepreneurship across diverse sectors. FinTech automates manual processes, reducing the resources and time needed for tasks such as loan processing and account opening, which saves costs (Khan et al., 2024). By digitizing processes and leveraging technology, operational costs are reduced by optimizing resource allocation and minimizing the need for physical branches. Online-only banks (neobanks) like Revolut, N26 and Chime are on the rise, offering enhanced accessibility and low-cost, user-friendly alternatives to traditional banking.

FinTech has enabled the development of user-friendly digital banking platforms and mobile apps, allowing customers to manage their accounts, transfer funds, and make payments conveniently from their devices (Rizvi et al., 2024). FinTech platforms facilitate peer-to-peer (P2P) lending, allowing individuals and businesses to borrow and lend money directly, often at more favourable interest rates than traditional banks. This is decentralizing the financial sector (DeFi). Advanced data analytics and AI furthermore enable banks to offer personalized financial products and services tailored to individual customer needs. In terms of improved operational efficiency, FinTech solutions, such as Robotic Process Automation (RPA) and machine learning algorithms, automate repetitive tasks like data entry and document verification (Hasan et al., 2024). This is reducing manual errors and optimizing resource utilization. Without a doubt, the digitalization of financial processes cuts costs and improves the speed and security of transactions, which enhances overall operational efficiency. In terms of risk management, FinTech solutions offer real-time monitoring and predictive analytics to help banks identify and mitigate risks more effectively. In promoting financial inclusion, FinTech innovations like mobile banking and digital payment systems have expanded access to financial services for previously unbanked or underbanked populations, particularly in developing regions (Sant'Anna et al., 2024). Additionally, FinTech platforms provide alternative financing options (e.g., crowdfunding and microfinance) for small businesses and start-ups, thus promoting entrepreneurship and economic growth.

Blockchain technology offers banks an opportunity to streamline operations and reduce fraud risks. Banks have started forming strategic alliances with FinTech companies to enhance their technological capabilities and service offerings (Karim & Lucey, 2024). This is expanding their customer base and improving customer satisfaction. However, FinTech also raises challenges around cybersecurity, regulatory complexities, increased competition, and financial inclusion. FinTech platforms store and process vast amounts of financial data, making them prime targets for hackers. FinTech-powered services have caused new cybersecurity concerns, especially for banks and businesses. Digital threats targeting FinTech services are deemed as highly dangerous weapons. For example, the 2022 Ronin Network hack resulted in a US$600 million cryptocurrency theft, highlighting the security risks in FinTech (Mohamad & Dimitriou, 2024). By potentially increasing the digital divide, digital banking innovations can be perceived as overly complex, compounded by a lack of digital and financial literacy. This could neutralize FinTech's efforts to improve financial inclusion. In terms of regulatory compliance, many FinTech start-ups operate outside traditional banking regulations, leading to consumer protection concerns. This is why regulatory technology (RegTech) solutions have been implemented to help banks comply with regulatory requirements more efficiently through automated reporting, real-time monitoring, and advanced analytics. According to De Koker et al. (2019), RegTech is improving the performance of customer due diligence, risk assessments, and transaction monitoring. RegTech also allows supervisory bodies to collect and analyse more provider data faster and more accurately, enabling them to intervene more quickly when compliance failures occur or risk rises to unacceptable levels.

Secondly, cryptocurrencies, blockchain technology, and decentralized finance (DeFi) are transforming the global financial system by introducing new models of value exchange, decentralization, and financial inclusion. These technologies challenge traditional banking, payments, and financial infrastructure, creating new opportunities and risks. Cryptocurrencies are digital or virtual currencies that use cryptography for security and operate on decentralized networks based on blockchain technology. Cryptocurrencies, like Bitcoin and Ethereum, enable decentralized, P2P payment systems, which is attracting attention from businesses, consumers, and investors (Javaheri et al., 2024). These new currencies are reshaping finance by enabling borderless transactions, digital asset investments, and alternative financial systems. They are also seen as a hedge against inflation.

Bitcoin, in particular, is increasingly viewed as a store of value in volatile economies. Cryptocurrencies are, however, introducing new risks and vulnerabilities, including crypto-mining and crypto-jacking attacks, along with uncertainty in the cryptocurrency market (Rizvi et al., 2024). Blockchain, as a disruptive technology, is being leveraged across various domains including FinTech, IoT, and the token economy.[3] Being a distributed ledger technology, it has a highly promising role in future information systems designs. As the backbone of financial innovation, blockchain promotes data privacy and transaction integrity in financial applications by separating transactional data from personally identifiable information, which is reducing privacy and security concerns associated with FinTech (Javaheri et al., 2024). Blockchain also powers DeFi. The latter is reinventing financial services. DeFi leverages blockchain technology and smart contracts to

[3] A token economy is the economic system built around a specific cryptocurrency or digital token. It is the model of how a crypto token will be used within a specific platform or ecosystem, and how value will be derived from it.

automate financial processes, which is reducing/eliminating the need for traditional intermediaries like banks, brokers, and clearinghouses (Karim & Lucey, 2024). DeFi platforms are generally open and permissionless, meaning anyone with an internet connection can access them. Examples include: (1) decentralized lending and borrowing platforms (e.g., Compound, MakerDAO, and Aave) that allow users to lend and borrow assets without traditional credit checks; (2) decentralized exchanges (DEXs) where platforms enable P2P trading of digital assets; and (3) 'yield farming', involving strategies that allow users to earn returns on their digital assets (Rajan, 2025). Perhaps the greatest benefit of DeFi is that it enables the creation of new financial products and services like synthetic assets, prediction markets, and decentralized insurance.

Thirdly, financial innovation in green finance and sustainability is reshaping global finance by directing capital towards environmentally and socially responsible operations and integrating ESG factors into investment decisions (Galeone et al., 2024). As the world shifts toward sustainability and climate action, green finance is channelling capital toward climate-friendly, socially responsible, and sustainable economic activities. Governments, financial institutions, and private investors are increasingly integrating sustainability into financial markets, risk assessment, and investment strategies. Here is how green finance is introducing a new paradigm in global finance:

- *Green bonds and sustainability-linked bonds*: These are fixed-income instruments specifically earmarked to raise money for climate and environmental projects (Hasan et al., 2024). The proceeds from green bonds are used to fund renewable energy projects, energy efficiency upgrades, sustainable water management, and other environmentally beneficial initiatives. Unlike green bonds, sustainability-linked bonds are not restricted to specific projects. Instead, they are linked to the issuer's overall sustainability performance, with financial terms tied to the achievement of predefined ESG targets.
- *Carbon trading and offsetting*: Financial markets for carbon credits allow companies to buy and sell permits to emit carbon dioxide (Pang et al., 2024). This creates a financial incentive for companies to reduce their carbon footprint. Companies can invest in environmental projects that reduce or capture greenhouse gas emissions to offset their own emissions. This includes reforestation projects, renewable energy installations, and methane capture initiatives.
- *Renewable energy financing*: Innovative financing structures are being used to fund large-scale renewable energy projects, such as solar farms, wind turbines, and hydroelectric plants (Pandey et al., 2023). These projects often involve public–private partnerships and risk-sharing mechanisms. Crowdfunding platforms like Mosaic and Abundance allow individuals to invest in renewable energy projects. This is democratizing access to green investments.
- *Sustainable insurance (InsurTech)*: Insurers are developing products that help businesses and communities manage the financial risks associated with climate change, such as extreme weather events and rising sea levels (Cosma & Rimo,

2024). There is also parametric insurance, which pays out based on predefined parameters (e.g., rainfall levels or wind speeds) rather than actual losses, providing quicker payouts and reducing administrative costs.

- *Green FinTech for sustainability*: FinTech companies are developing platforms that provide real-time ESG data and analytics, helping investors make informed decisions based on sustainability criteria. Blockchain technology is being used to enhance transparency and traceability in supply chains, ensuring that products meet sustainability standards (Pang et al., 2024). In this way, blockchain is integrated into financial services to promote green financing and green innovation programs, leading to green growth.
- *Innovative financial instruments*: Climate bonds are specifically designed to finance projects that address climate change, such as renewable energy, energy efficiency, and sustainable transportation (Van Niekerk, 2024). Similar to green bonds, there also social bonds that raise funds for projects with positive social outcomes, such as affordable housing, healthcare, and education.

Banks are developing strategies to improve the environmental and social impact of their investment and financing choices to achieve the UN's SDGs. The implementation of financial technology has proved necessary for facilitating the transition to a sustainable economy faster and more efficiently (Galeone et al., 2024). Financial innovation in green finance and sustainability is integrating environmental and social considerations into financial decision-making and it is integrating climate action into capital markets, banking, and investment strategies. Innovations in ESG investing, carbon markets, and blockchain-powered climate finance are driving the transition to a more sustainable, low-carbon and resilient global financial system, creating new opportunities for investors, businesses and communities.

As the world continues to face pressing environmental and social challenges, the role of green finance in promoting sustainable development will only become more critical. However, challenges remain, including greenwashing,[4] regulatory inconsistencies, and the need for greater financial inclusion in climate finance (Sant'Anna et al., 2024). A balance needs to be found between enabling financial innovation and addressing challenges and risks to financial integrity, consumers' protection, and financial stability. Ultimately, green finance is no longer an option—it is the future of global finance.

2.5 Economic Sustainability

The prioritization of economic sustainability is notably influencing the trajectory of the global economy in the twenty-first century, with an increasing emphasis

[4] Greenwashing is when a company misleads consumers by falsely claiming or exaggerating its environmental efforts to appear more sustainable than it actually is. This can involve deceptive marketing, vague claims, or misleading labels to attract eco-conscious customers while continuing environmentally harmful practices in secret.

on sustainable practices and a move away from traditional economic models. In particular, ESG considerations are integrated into economic policies, business practices, financial markets, and investment decisions (Cricelli & Strazzullo, 2021). This shift is driven by the recognition that long-term economic growth and stability depend on sustainable practices that balance economic development with environmental protection and social equity. To put it another way, it is driven by a growing appreciation of the interconnectedness between economic prosperity, environmental health, and social well-being.

The SDGs are arguably making the most significant impact as far as steering the global economy towards economic sustainability. These goals provide a comprehensive framework for achieving economic sustainability. The 17 goals address critical issues such as poverty, inequality, climate change, and environmental degradation. Governments and international organizations are integrating the SDGs into their economic policies and development plans, ensuring that sustainability is a core consideration in economic decision-making. Businesses are also starting to integrate the SDGs and ESG factors into their operations, investment decisions, and reporting. This reflects a recognition that sustainability is not just an ethical imperative, but also a key driver of long-term value creation. For example, in Bangladesh, the Grameen Bank's microfinance initiatives have empowered millions of low-income individuals, particularly women, to start small businesses and improve their economic conditions. Sustainable development now intertwines economic development, environmental protection, and social equity (Nikolaou et al., 2021). This integration acknowledges that long-term economic growth must preserve environmental and social resources. Other key trends steering the global economy towards economic sustainability, include:

- *Transition to a green economy and a circular economy*: The green economy is emerging as a framework for climate mitigation, economic growth, and poverty alleviation, with key investments directed towards increasing natural capital and energy and resource productivity (Alavijeh et al., 2017). The transition to a green economy involves investing in renewable energy, energy efficiency, sustainable agriculture, and green infrastructure. This shift is creating new industries and job opportunities while reducing environmental impact. The circular economy model emphasizes reducing waste, reusing materials, and recycling resources. This approach minimizes resource extraction and waste generation, promoting sustainable consumption and production patterns.
- *Regulatory and policy support*: Governments and regulatory bodies are developing classification systems—called green taxonomies—to define what constitutes a sustainable investment. This is helping to standardize and legitimize green finance practices. The implementation of carbon pricing mechanisms, such as carbon taxes and cap-and-trade systems, incentivizes businesses to reduce their carbon emissions and invest in cleaner technologies. Regulations requiring companies to disclose their ESG performance and adhere to sustainability standards

are increasing transparency and accountability. The EU, for example, is focusing on the digitalization of industry through technologies like big data and AI to achieve environmental and economic benefits (Cricelli & Strazzullo, 2021).

- *Growing emphasis on sustainable consumption*: Increasing consumer awareness of the environmental and social impact of their purchasing decisions is leading to an increased demand for sustainable products and services. This is prompting businesses to adopt more responsible practices (Hasan et al., 2024). It is reducing waste and lowering production costs, which enhances long-term economic resilience. Increased consumer demand for eco-friendly products is pushing businesses to invest in sustainable technologies, circular economy models, and renewable energy, which is leading to new economic opportunities. Consumers now have easier access to information regarding the environmental impact of their consumption choices. Digital tools, like smart packaging with QR codes, give consumers more product information, including sourcing and production details, which assists them in making sustainable and ethical purchasing decisions.
- *Smart technology for sustainability*: Advances in clean technologies, such as renewable energy, energy storage, and electric vehicles, are driving the transition to a low-carbon economy. Biotechnology and renewable energy innovations are now playing a central role in shaping sustainable industries. Digital technologies and FinTech are identified as catalysts for advancing sustainability (Galeone et al., 2024). They enable streamlined financing for sustainable practices, transparent supply chains, and real-time data analytics for informed decisions. Unlike in the past, sustainability is now driving technological innovation in areas such as renewable energy, energy efficiency, and waste management. The use of AI, blockchain, and IoT to optimize energy use, reduce waste, and track carbon footprints is increasing. Smart grids are also being developed as part of sustainable urban planning.
- *Global cooperation and partnerships*: Global agreements, such as the UN SDGs and the Paris Agreement on climate change, foster international collaboration on sustainability goals and it encourages countries to adopt ambitious climate targets and policies. The IMF, World Bank, and WTO are all pushing for sustainable development policies in member nations. There are growing efforts towards collaboration between governments, businesses, and civil society; these are essential for scaling up sustainable solutions and achieving the SDGs (Van Niekerk, 2020). Through such public–private partnerships, new sustainable economic models can be tested and developed. Due to a growing recognition of the limits to economic growth, nations are starting to assist each other in developing steady-state economies. This involves reconfiguring international trade and identifying alternative paths for developing economies to increase their citizens' well-being.

It is important to note that the digital economy, while offering opportunities for sustainability, also presents challenges such as rising e-waste, high energy consumption, and data protection problems (Rosário & Dias, 2023). For economic inclusivity to reduce global inequality, it requires full integration into business models, public policy, and community development. Another challenge to economic sustainability is the fact that there can be inconsistencies in implementation, such as environmental laws not being enforced due to pragmatic economic interests. Currently, the risk of trade wars, financial bifurcation, and technological decoupling threatens global economic stability and the achievement of SDGs. At least Industry 4.0 focuses heavily on sustainable development, with manufacturing incorporating 'lean and green' production approaches. Combined with sustainable energy integration, it is essential for accomplishing both global climate mitigation and sustainable development goals (Pandey et al., 2023). Finally, to bring lasting change, significant effort in planting the seeds of economic sustainability is required at grassroots level. Schools and universities should prepare future generations to work in the emerging economy by aligning their skills with trends in the digital sphere (Nyagadza et al., 2022). Societies in emerging economies will need to prioritize retraining workers whose skills have been timed out.

2.6 A New Wave of Economic Nationalism

Economic nationalism is the prioritization of domestic industries, jobs, and national interests over global economic integration. It is an ideology that emphasizes state intervention in the economy through policies like domestic control, tariffs, and restrictions on the movement of labour, products, and capital. At its core, it asserts that economic policy should align with nationalist objectives (Helleiner, 2021). As a dominant modern ideology, economic nationalism—in its classical sense—contrasts with both economic liberalism and economic socialism. Proponents of economic nationalism resist globalization and often challenge the benefits of unrestricted free trade. It is nationalists vs. globalists.

Economic nationalists believe that there are limits to globalization, which requires state intervention. They advocate protectionism and prioritize national self-sufficiency. In this framework, markets are subordinate to the state and are expected to serve strategic national interests, such as ensuring national security and bolstering military strength. Economic nationalists typically view international trade as a zero-sum game, where the focus is on securing relative advantages rather than mutual benefits (Nakano, 2007). The ideology also strongly promotes industrialization, often supporting and subsidizing domestic industries through state intervention. This preference stems from the belief that industrial growth fosters broader economic benefits, strengthens national autonomy, and plays a vital role in military development.

It is therefore fair to ask whether economic nationalism has any role to play in an increasingly interdependent global economy. The meteoric rise of Trumpian capitalism during the past decade serves as a reminder that as much as progressive globalism have become dominant in the late 1990s and early 2000s, it also served to stir up opposition forces. Trumpian capitalism is the latest version of such a force in the form of 'new' economic nationalism (Mollan & Geesin,

2020). Given Trump's growing influence in the world through being the President of the United States of America, his administration's influence and the policies they set is having a tremendous impact on how the global economy is evolving in the twenty-first century. Although it is now in a different (4IR) context, the emergence of economic nationalism is an old-new phenomenon in the global economy (Jędrzejowska, 2017). The GFC of 2007–09 was perhaps the turning point, away from hyper-globalization towards protecting national interests. While some initially predicted a generalized rise in protectionist barriers after this huge financial crash, this has not fully materialized in the form of a complete withdrawal from global markets (Gaudiosi, 2018). However, it is undeniable that since then, a new wave of economic nationalism is evident in the policies of major global players. Not just the US, but Europe—after the Euro crisis in 2010—have implemented a number of nationalist policies. Apart from stronger financial regulation, policies that protect the European labour market and policies dealing with issues related to opening the market to agricultural products from outside the EU, are examples of defensive economic nationalism at both the member state and EU levels (Jędrzejowska, 2017). Negotiations on agreements like the Transatlantic Trade and Investment Partnership (TTIP) also revealed such tendencies. Brexit is another example of a move towards economic nationalism. The UK's departure from the EU has changed its trade relationship with the EU fundamentally, which previously accounted for a substantial portion of its exports. This signals a move towards greater national control over trade policy. These policy shifts and change in momentum have led to the resurgence of nationalist parties in a number of countries within the EU, which has entrenched the focus on national interests. Globally, the COVID-19 pandemic has certainly played a significant role in prompting nations to prioritize national needs and self-reliance, pushing countries further into a nationalist economic turn (Rewizorski, 2021).

There are several ways in which economic nationalism is currently impacting and transforming the global economy. Firstly, in terms of trade policies and protectionism, a variety of measures aimed at prioritizing national economic interests over global free trade have been, and is being, implemented in countries around the world. Probably the most widely publicized current development is the US, under the new Trump administration, raising exceptionally high tariffs on imports from trading partners like Canada, Mexico, Europe, and China. The US-China trade war is now escalating to unprecedented levels after the imposition of tariffs on billions of dollars' worth of goods, leading to increased costs for businesses and consumers on both sides (LaRocco, 2025).

There is also an increase in local content requirements, as governments require that a certain percentage of goods or services be produced domestically to promote local industries, but potentially limiting foreign investment and competition. For example, Brazil's local content requirements in the oil and gas sector have aimed to stimulate domestic manufacturing and job creation (Pereira et al., 2023). The US's Inflation Reduction Act (IRA) and the EU's Green Industrial Plan are further examples of favouring domestic production of clean energy technologies over imports. Nations are emphasizing self-reliance in critical industries like semiconductors, pharmaceuticals, and energy, and they are reducing dependence on foreign production by reshoring (bringing production back home, which is import substitution) or near-shoring (moving supply chains to nearby allied nations) (Capello & Dellisanti, 2024).

Secondly, countries are using industrial policy and self-sufficiency to protect and strengthen themselves. Governments are increasing investment in strategic sectors, providing subsidies and incentives to support domestic manufacturing and innovation. Economic nationalism emphasizes the importance of strategic industries, such as technology, defence, and healthcare, for national security and economic independence (Rammal et al., 2022). In this way, industrial policies favour local producers over foreign competitors. This traditionally focused on manufacturing, but now includes providing stimulus for innovation and technological advancement. Examples include the European Union's efforts to develop a domestic semiconductor industry to reduce reliance on foreign suppliers and enhance technological sovereignty; China's 'Made in China 2025' strategy aims to dominate key high-tech industries; and The US CHIPS and Science Act seeks to reduce reliance on Asian semiconductor manufacturers. Economic nationalist policies often aim to foster 'national champions'—specific firms or sectors deemed crucial for national competitiveness and development (Pryke, 2012). Driven by a desire to not have a nation's fortunes determined solely by world markets, nationalist policies also aims to achieve greater economic self-sufficiency, especially in critical sectors like agriculture and heavy industry. While fostering national interests, some forms of modern economic nationalism involve strategically leveraging globalization to benefit domestic firms (Rammal et al., 2022). This means utilizing local resources to extract maximum benefit from the global economy rather than simply protectionism. States might pursue policies that attract FDI in specific sectors deemed beneficial for national development.

Thirdly, countries use national security and technology as a safeguard. The drive to prioritize national economic interests increasingly intertwines with concerns about security and the desire for technological advancement and autonomy. Economic nationalism often involves restrictions on the transfer of sensitive technologies to foreign entities, particularly in sectors like defence, telecommunications, and AI. For example, the US has imposed restrictions on Chinese tech companies like Huawei and ZTE, citing national security concerns and the need to protect critical infrastructure (Erie & Streinz, 2021). Governments are also increasingly focused on protecting national data and ensuring that critical digital infrastructure is controlled domestically. One of the best examples is the EU's GDPR, which emphasizes data sovereignty and the protection of citizens' personal data (Skare & Soriano, 2021). This shows how economic nationalism fuels techno-nationalism, where nations prioritize the development and control of key technologies. States increasingly find it necessary to intervene in science and technology through R&D funding, institutional mechanisms, and support for private initiatives to build competitive knowledge-intensive sectors (Diesen, 2017). This is seen as essential for national security and future economic prosperity. Russia's funding of Skolkovo and adoption of import substitution in sectors like the airline industry, as well as China's focus on technological progress in its economy, exemplify this.

Economic nationalism is fostering a global environment where national security and technological advancement are seen as integral to economic strength and sovereignty. This is leading to more interventionist industrial policies focused on nurturing domestic technological capabilities, protecting strategic industries, and controlling critical technologies and data flows, often at the expense of unfettered global exchange and cooperation. Other drivers of economic nationalism include:

- *Foreign investment and capital controls*: Countries are implementing stricter screening and restrictions on FDI to protect national interests and prevent foreign control of key industries. For example, the Committee on Foreign Investment in the US (CFIUS) has increased scrutiny of foreign investments, particularly from China, to safeguard national security (Jalinous & Sensenig, 2022). Countries have become more sensitized to implementing capital controls to manage the flow of foreign capital and protect domestic financial stability. India, for example, periodically use capital controls to manage volatile capital flows and stabilize its currency.
- *Public procurement and SOEs*: Governments are increasingly implementing 'buy local' policies for public procurement, favouring domestic suppliers over foreign ones (e.g., Proudly South African and Buy American). This is driven by a culture of consumer nationalism that governments seek to promote to direct public spending towards local businesses. Furthermore, economic nationalism often involves the use of SOEs to control key sectors of the economy and ensure national interests are prioritized (Kalu & Ogbonnaya, 2019). For instance, China's SOEs play a strong role in sectors like telecommunications, transportation, and energy—in line with its national economic goals. Economic nationalism leads to a greater scrutiny and often resistance towards foreign takeovers of domestic firms, especially those considered strategic or state-owned.
- *Impact on international cooperation*: The rise of economic nationalism can impact global cooperation efforts on global issues like climate change and social inequality negatively, as nations give preference to their own economic interests. The trend of large nations like the US withdrawing from international agreements also reflects this. Economic nationalism can also be a challenge to multilateral institutions like the WTO and the IMF to perform their role effectively, as countries prioritize national interests over global cooperation. The US has, for instance, criticized the WTO for not adequately addressing issues like intellectual property theft and unfair trade practices, leading to calls for reform.
- *New forms of economic nationalism*: Contemporary economic nationalism is not solely reliant on traditional protectionist tools like tariffs. It manifests in more nuanced ways like 'long-distance free market nationalism' driven by diasporas, 'financial nationalism' using monetary and fiscal policies for nationalist agendas, and a focus on promoting national brands and discouraging foreign brands

(Mikecz, 2019). Financial nationalism can involve increasing state ownership in strategic financial sectors or implementing specific financial regulations to favour national interests. Green nationalism or eco-nationalism is another form that emerged as some countries are using economic policies to achieve environmental goals, often with protectionist measures (Obstfeld, 2021). Examples include the EU's Carbon Border Adjustment Mechanism (CBAM) for tariffs on imports from countries with weaker climate policies; the US's IRA, which allows subsidies for domestic green energy production and electric vehicles; and China's dominance in green tech whereby they are securing global supply chains for batteries, solar panels, and rare earths.

Economic nationalism is a significant force in reshaping the twenty-first century global economy. Despite coexisting in a complex relationship with globalization, its resurgence is leading to new forms of state intervention, protectionist measures, and a focus on national economic interests and security. While it can drive innovation, protect jobs, and enhance national resilience, it also poses challenges to global economic integration, trade, and cooperation. The balance between protecting national interests and fostering global collaboration will be crucial in determining the global economy's future trajectory. As countries navigate the complexities of economic nationalism, the need for strategic policies that promote both domestic growth and international stability will remain vital.

2.7 BRICS+ and De-dollarization

The BRICS countries have emerged as a powerful force in the global economy, with the potential of changing the current world order substantially. The economies represented in the BRICS alliance have sustained strong growth in recent decades, of which four have annual GDPs exceeding US$1 trillion. The combined GDP of the BRICS reached US$17 trillion in 2014, representing just below 22% of the global economy. By 2023, BRICS countries accounted for 32.14% of the global GDP in terms of PPP, surpassing the G7 nations at 29.92% (Nach & Ncwadi, 2024). The BRICS countries hold significant foreign currency reserves, accounting for around 40% of the world's total. The nations are considered attractive destinations for FDI and have also grown as investors in their regions. BRICS countries account for a large percentage of the world's population (40%) and land coverage (30%). This will now grow even more with the addition of Iran, Egypt, UAE, Ethiopia, and other possible members, confirming the growing significance of BRICS+ in the global economy. This expansion aims to further challenge existing global power structures and advance de-dollarization trends (Gouvea & Gutierrez, 2023). BRICS+ represents a significant geopolitical development focused on reshaping the global financial and economic landscape, promoting economic sovereignty and South-South cooperation.

As pointed out in Chap. 1, BRICS+ is undoubtedly challenging the Western-dominated global order and reshaping it into a multipolar world order. China, as the member with the largest economy in the alliance, envisions BRICS as a strategic platform to advance its geopolitical objectives, including promoting initiatives such as the BRI and adopting a more overtly oppositional stance toward the United States (Walker, 2024). In its venture to create a more multipolar world order, BRICS+ specifically wants to reduce the dominance of Western-led institutions like the IMF and World Bank. They advocate for greater voting power and representation of developing countries in global governance and are constantly pushing for reforms in global institutions to better reflect the current economic realities. This diversification would reduce the relative economic dominance of traditional powers like the US and the EU.

The BRICS+ nations are increasingly vocal in international forums, advocating for policies that support sustainable development and equitable growth (Shameem & Jayaprasad, 2020). Their vision seeks a balance between the North and South in global politics. By shifting the economic centre, BRICS plays an important political role as a counterpart to the US-EU-western dominance model, focusing on partnership rather than a donor-recipient relationship. The BRICS countries have expressed a strong desire to participate in and act on global economic governance, supporting multilateralism and promoting domestic reform to make the system more equitable (Vyas-Doorgapersad, 2022). Hence, they have embarked on institutional innovation by creating alternative institutions that better represent emerging economies. The NDB and the AIIB provide infrastructure financing and funding for sustainable development projects without the political conditions often imposed by the IMF and World Bank. To the BRICS+ countries, these Eastern institutions provide a better balance in economic policy, trade agreements, and financial systems. Russia and China lead efforts to challenge Western-led economic frameworks, while India, Brazil, and South Africa advocate for more inclusive global institutions. As a growing group of nations, BRICS+ fosters economic cooperation and trade among developing countries, creating new markets and opportunities. This is strengthening economic ties between nations in the Global South, making them more globally competitive.

By representing a growing portion of global trade, the expansion of the BRICS alliance brings in more nations that have large amounts of natural resources, so changing global trade patterns. BRICS+ countries are both significant sources and destinations for FDI, contributing to global capital flows and investment patterns. Manufacturing hubs are shifting as companies adopt 'China + 1' strategies, moving production to Vietnam, India, and Mexico to mitigate geopolitical risks. With this growing economic power, the BRICS+ countries are actively involved in the G20, using it as an important venue to shape global economic governance and call for enhanced representation of rising markets (Moch, 2024). These developments reveal the BRICS+ goal of forming a complementary financial system tailored to favour developing nations. With the US losing its grip, countries are seeking currency alternatives amid global sanctions and new Global South alliances. Although the dollar is still recognized as the world reserve currency, countries around the world are now openly discussing the creation of alternative methods to conduct trade and settlement, as well as reducing their dollar reserves (Giustra, 2023).

BRICS+ is playing a central role in this probable transition as countries have been aggressively adding to their gold reserves after the GFC, with the size of their purchases increasing over time, suggesting that any new currency would in all probability be backed by gold.

The list of alternatives to the US dollar is expanding rapidly. Notable examples include China experimenting with cross-border digital currency settlements with Thailand and the UAE, as well as requiring sanctioned nations like Russia, Iran and Venezuela to accept yuan as payment for oil. Saudi Arabia is reportedly considering a similar move, with speculation that it is already selling oil for yuan and converting those yuan into gold on the Shanghai exchange (Nieuwenhuijs, 2023). Additionally, India has begun purchasing Russian oil in UAE dirhams. A growing trend is the use of bilateral agreements that facilitate trade in local currencies. A sharp decline in demand for the US dollar could have severe repercussions for Americans. It might spark a dollar crisis, resulting in extreme inflation—or even hyperinflation—and set off a cycle of escalating debt and excessive money printing, potentially destabilizing the social fabric of the nation. Hence, the new Trump administration considers any such de-dollarization efforts to be matters of national security. But as Giustra (2023: 1) states, "de-dollarization is not a matter of if, but when". Regardless of US resistance, de-dollarization is expected to continue, as much of the non-Western world seeks a trading system that shields them from the risks of dollar dominance and financial coercion.

BRICS+ nations are strategically focusing on de-dollarization and working toward reducing reliance on the US dollar in trade and finance. This is weakening the global influence of the dollar. Many other emerging economies are doing the same in trade settlements, favouring local currencies, gold-backed settlements, and CBDCs (Liu & Papa, 2022). US sanctions and financial restrictions have pushed China, Russia and other nations to promote alternative payment systems like the digital yuan and BRICS-led initiatives. China's digital yuan and initiatives like BRICS payment systems are challenging the SWIFT-based global financial order. Other alternative financial mechanisms include:

- the BRICS Contingent Reserve Arrangement (CRA), which acts as a mini-IMF to help member countries deal with currency fluctuations;
- China's Cross-Border Interbank Payment System (CIPS) competes with SWIFT, helping BRICS nations bypass Western financial controls; and
- discussions are ongoing about creating a BRICS digital currency or a commodity-backed alternative to the dollar (Mosharrafa, 2024).

China, in particular, is stimulating trade within the BRICS framework, in anticipation of at some point moving away from the US dollar. China has also signed agreements with numerous nations for settling bilateral trade in yuan (Aggarwal, 2020). As another critical step, BRICS+ countries are discussing and planning for the establishment of a new Multilateral Clearing Union (MCU) within the framework of the NDB to serve as an intra-BRICS currency swap pool and de-dollarize BRICS trade.

The BRICS+ nations are increasingly playing a central role in the direction of the global economy through their significant economic growth, the establishment of alternative financial institutions, efforts towards de-dollarization, economic sovereignty, and their increasing political influence in international forums. While internal disputes and China's dominant role exist within the alliance, these countries—collectively—strive for a more equitable and multipolar global economic order (Larionova & Shelepov, 2022). The expansion to BRICS Plus further amplifies their potential impact on the global economic landscape. BRICS+ represents a major force, making its growing influence undeniable. The collective impact of BRICS+ is transforming the global economy by the day as these nations are driving growth, shifting economic power, enhancing trade and investment, fostering innovation, developing infrastructure, and increasing their geopolitical influence.

3 What Is an Inclusive Economy and Why Does It Matter?

Before answering the questions above, let's first consider what is meant by economic inclusion. It is the process of ensuring that all individuals, especially those from disadvantaged or marginalized groups, have equal access to economic opportunities, financial services, and resources necessary to participate fully in the economy (Van Niekerk, 2022). It aims to reduce inequality by integrating more people into the financial system, the labour market, and entrepreneurship to stimulate productive participation. It is not just inclusion for the sake of inclusion (free riding), but productive inclusion, to ensure that value is being added to the economy. In this way, not only the marginalized but also those already contributing to the economy, benefit from the additional growth, income, and opportunities created (Schoneveld, 2020). This point is important, because it is often mistakenly assumed that economic inclusion is only about helping the poor. The real benefit of economic equity is how it improves everyone's quality of life in the broader society from short—to long term. Here are ten benefits, among others, of economic inclusion:

- *Economic growth and expansion*: Including marginalized groups expands the consumer base, leading to higher demand for goods and services. This can drive business growth and create new market opportunities. A more inclusive economy can lead to higher overall economic output and GDP growth, benefiting everyone through increased economic activity.
- *Enhanced productivity and innovation*: Including marginalized groups brings diverse perspectives and talents into the workforce, fostering innovation and creativity. Leveraging the skills and abilities of a broader population can lead to more efficient and effective production processes.
- *Reduced social costs*: Reducing poverty through economic inclusion can decrease the social costs associated with poverty, such as crime, healthcare expenses, and social welfare programs. Better economic opportunities for

marginalized groups can lead to improved public health outcomes, reducing the burden on healthcare systems. There is also a positive correlation between people being healthy and productive (Patel et al., 2022).

- *Social stability and cohesion*: Lowering economic disparities quite often lead to greater social stability and reduced social tensions, creating a more harmonious society (Muntaner et al., 2020). Inclusive economic policies can lead to stronger, more resilient communities, where collaboration and mutual support flourish. All residents benefit from this as well as from new opportunities to pool resources for improving collective well-being.
- *Increased tax revenues*: As more people gain employment and earn higher incomes, the tax base expands, leading to increased public revenues and perhaps lower tax rates. These revenues can be used to fund public services and infrastructure that benefit everyone. With more people employed and self-sufficient, there is less reliance on social welfare programs, freeing up resources for other public investments.
- *Enhanced corporate reputation and brand loyalty*: Companies that actively promote inclusion and diversity can enhance their reputation and build stronger brand loyalty among consumers who value social and ecological responsibility. Inclusive workplaces often have higher employee satisfaction and retention rates, reducing turnover costs and improving overall productivity (Ciucu et al., 2024).
- *Access to new markets and opportunities*: Marginalized groups often bring unique insights and ideas, leading to the creation of new businesses and markets that can benefit the broader economy. A diverse and inclusive workforce can enhance a country's competitiveness in the global market by fostering innovation and adaptability. In addition, efforts to recycle and upcycle waste can be turned into new entrepreneurial ventures or added to businesses' value chains (Raworth, 2018).
- *Strengthened social networks*: Economic inclusion can strengthen social networks and community ties (both physical and online communities), leading to more cohesive and supportive communities. Inclusive economies often foster collaboration between different sectors and groups, leading to new partnerships and opportunities for growth.
- *Long-term sustainability*: An inclusive economy is more resilient to shocks and stresses, as it draws on the strengths and resources of a broader population. This is the power of collectivism and Ubuntu.[5] Inclusive economic practices and policies align with the SDGs, ensuring long-term environmental, social and economic sustainability.

---

[5] Ubuntu is a fundamental African philosophy and principle that accentuates interconnectedness, community, and humanness towards others. It signifies a relational worldview, emphasizing collective progress and accountability.

- *Moral and ethical benefits*: Promoting economic inclusion aligns with principles of social justice and fairness, contributing to a more equitable, responsible, and just society. Ensuring that everyone has the opportunity to contribute to and benefit from economic growth fosters a sense of shared prosperity, accountability, and collective well-being.

Economic inclusion is a basic condition for creating a sustainable and more diversified economy. It implies a fundamental reorganizing of society where values and processes include rather than exclude (Fioramonti, 2017). It aims at including all (willing and able) members of society in the growth process itself instead of only distributing wealth among them after periods of steep growth. Economic inclusion is not just about policies that prioritize equal opportunities, economic safety nets, and pro-poor strategies, but also about embedding inclusivity into every part of economic life organically. This requires political will, inclusive institutions, and policies that are designed to work for everyone, rather than being mere add-ons to existing systems.

Additionally, financial inclusion plays a key role in economic inclusion by providing access to banking, credit, insurance, and digital financial services, particularly for the unbanked and underserved populations (Ozili, 2021). Inclusive banking initiatives, such as microfinance, community banks, and digital financial tools, contribute to stimulating local economic development and broadening participation in the economy. Economic inclusion is essential for achieving sustainable and equitable economic growth. By ensuring that all individuals and communities have the opportunity to participate in and benefit from economic activities, societies can reduce poverty, inequality, and social exclusion, leading to more stable and prosperous communities.

Now we can ask, what is an inclusive economy? It is an economy driven by inclusive growth with the goal of genuine economic progress to promote equality of opportunity and broader well-being for both people and the planet (Van Niekerk, 2022). The primary goal is to reduce disparities and create a more equitable economic environment where everyone, regardless of their background, can add value to the economy. The secondary goal is to prevent social exclusion and economic crises. By re-emphasizing the original word for economics (the Greek word, *oikonomos*, meaning household management), the inclusive economy reclaims the economy's original purpose: good stewardship of resources. To show the contrast, Aristotle identified *chrematistics*, which is an antithesis to *oikonomos*, meaning the manipulation of property and wealth/money to maximise an owner's short-term monetary exchange value (Kraut, 2006). This normally results in greed, exploitation, and pervasive externalities. The latter is what is making the current supercapitalism model unsustainable and the success of modern-day socialist economies (like China) so attractive. The inclusive economy takes elements of both capitalism and socialism, such as the market-based system and collectivism, but repositions them in a unique combination that ensures economic equity, environmental sustainability, and genuine progress in a way that protects people's individual freedom and promotes decentralization instead of centralization (a socialist tendency).

By emphasizing collective participation and benefit distribution, the inclusive economy represents a rebalancing[6] to ensure collective well-being. Essentially, it seeks to move beyond a growth-centered approach to shape an economy that benefits both individuals and the collective, while also safeguarding the planet and its resources (Van Niekerk, 2020). It fully embraces innovation as far as it is for the purpose of improving sustainability, not exploitation. In this regard, there are a number of core principles associated with an inclusive economy, which include:

- *Prioritizing people and the planet*: The foundation upon which economic systems are built and sustained are people and the planet. When they are valued properly in terms of their contribution to the economy (skills and resources), their value appreciates. Unlike financial capital, which is also important to the economy, human and natural capital thrive in the inclusive economy. The principle: never prioritize financial capital at the expense of human and natural capital; rather prioritize all three in holistic balance.
- *Equality of opportunity and broader well-being*: Economic growth should not be measured solely by metrics like GDP, but should also consider how opportunities and benefits are distributed across society and how well-being is enhanced for both current and future generations. Notably, the emphasis on human well-being (broader than GDP growth) is integral to the humanness philosophy of Ubuntu. The principle: economic systems should prioritize inclusivity, fairness, and sustainability to ensure long-term prosperity for all.
- *Preventing social exclusion and economic crises*: Addressing inequality, fostering inclusion, and building safeguards to mitigate the risks of economic downturns is critically important to any economy, given increased economic interdependence. An inclusive economy works to prevent social exclusion and it increases the economy's ability to absorb the shocks of financial crises. The principle: inclusive and resilient economic systems are essential for long-term stability, growth, and societal well-being.
- *Inherent inclusivity*: Originally, economic activity was seen as inherently inclusive, even as regards the community as a whole, stemming from the Greek word *oikonomos*. Over time, the focus increasingly turned toward profit-seeking and wealth accumulation, which contributed to rising exclusion and inequality. The inclusive economy brings the emphasis back to collective participation and benefit distribution. The principle: economic systems should be intentionally designed to be inclusive, ensuring equal access to opportunities, resources, and benefits from economic activities for all individuals and communities, rather

[6] Rebalancing in this context means, whereas the original focus of the economy was on collective participation in production and distribution of benefits, and it, later on, shifted towards money-making and fortune-building, resulting in rising exclusion and inequality; the inclusive economy now places people and the planet at the centre again.

than perpetuating barriers that favour a select few (entrenching exclusion and inequality).

It is important to ask: What makes the inclusive economy different from the classic, Western free- market economies, and different from the Eastern socialist economies? As can be seen from.

Table 1, each model has its strengths and weaknesses, but the inclusive economy seeks to combine the best aspects of market efficiency and social equity, while addressing the shortcomings of both neoliberalism and socialism. With its focus on fairness, equity, sustainability, and ensuring that everyone benefits from economic growth, it balances market efficiency with social welfare. In contrast, the market-led economy primarily focuses on economic growth through free markets and minimal government intervention, often emphasizing individual wealth and competition. Meanwhile, the socialist economy aims for social ownership and centralized control of production to achieve social and economic equality, prioritizing collective needs over individual profit.

So, what are the essential requirements or components that makes an economy inclusive? Or, put differently, when can an economy be considered as inclusive? An inclusive economy requires a fundamental reorientation of economic thinking and practice, moving beyond a sole focus on growth to prioritize the well-being of people and the planet through collaboration, ethical considerations, and systemic changes guided by principles like Ubuntu. An inclusive economy requires intentional design, proactive policies, and sustained commitment to ensure that everyone can participate in and benefit from economic activities. By addressing barriers to participation, reducing inequality, and prioritizing sustainability, societies can create economies that are not only more equitable but also more resilient and prosperous for all. Therefore, to answer the question, an economy is inclusive when it ensures that all individuals, communities, and businesses have equal opportunities to participate in and benefit from economic growth. It is also inclusive when it prioritizes equity, sustainability, and resilience, thereby ensuring that economic benefits are widely shared rather than concentrated among a few. To break it down, the following are minimum requirements for an inclusive economy:

- *Inclusive growth*: It promotes equal opportunities for all economic participants during the growth process (not just after periods of growth), ensuring that the benefits are incurred by every segment of society. For growth to be inclusive, it needs to be sustainable and effective in reducing poverty and respect planetary boundaries. GDP growth should be fully pursued, but there is more to be achieved. This is multi-dimensional/holistic progress that prioritizes equality of opportunity, broader well-being, and markedly reduces inequality. It combines pro-poor growth, broad-based growth, and shared growth.
- *Balancing economic progress with social equity and ecological parity*: Genuine economic progress is the net effect of subtracting negative externalities (like social and environmental costs) from GDP growth and adding positive externalities not measured by GDP. When economic inclusion promotes social equity,

Table 1 Comparing the inclusive economy with other economic models

Feature	Western free market economy	Inclusive economy	Eastern socialist economy
Core philosophy	Prioritizes free markets, deregulation, and private enterprise for maximum efficiency and growth	Balances economic growth with equity, sustainability, and broad participation in a free market plus Ubuntu values	Emphasizes state control over the economy, wealth redistribution, and collective welfare; power is centralized
Economic goal	Maximizing economic growth, efficiency, and individual wealth accumulation	Broad-based well-being, reducing inequality, ensuring access for all, raise quality of life, and genuine progress	Equal distribution of resources and elimination of class disparities; development through infrastructure
Role of the state	Minimal state intervention; protecting property rights, the market determines prices, wages, and competition	Supports regulation to ensure fairness, resource distribution, social mobility, environmental protection; state support market only as needed	Strong government intervention, with central planning of industries and resource allocation; state control over companies
Economic equality	Accepts income inequality as a natural outcome of competition and market efficiency	Aims to reduce inequality through inclusive policies, fair wages, community and stakeholder collaboration, and inclusive business ventures	Focuses on economic equality through state ownership and wealth redistribution. Limited success over history
Ownership structure	Primarily private ownership of businesses and assets, with some public services	Mix of private enterprises, cooperatives, and social enterprises with state oversight and support	State-owned enterprises dominate key industries, with some private ownership in select sectors
Welfare and social safety nets	Limited social safety nets; individuals must rely on market-driven solutions	Strong welfare programs (universal healthcare, education and housing) with a focus on accessibility	Comprehensive welfare system, often providing universal healthcare, education, and housing

(continued)

Table 1 (continued)

Feature	Western free market economy	Inclusive economy	Eastern socialist economy
Wealth distribution	Wealth accumulation is driven by market success; inequality is often high	Progressive taxation and social programs to reduce inequality; social networks collaborate	Wealth is redistributed by the state; aims for equality but may lack incentives
Labour rights and employment	Flexible labour market; wages and job security are dictated by supply and demand	Prioritizes decent work, living wages, entrepreneurship, and worker participation in economic decision-making	Government controls employment, wages, and worker protections through centralized planning
Sustainability, environment and business	Environmental policies are secondary to profit and economic expansion; minimal restrictions on business practices	Strong focus on sustainable development, green finance, climate-friendly policies, inclusive business (e.g., fair wages, diversity, circular)	State-driven environmental policies, but often prioritizing industrial growth over sustainability; private sector is limited, under state control
Financial and monetary system	Dominated by private banks and capital markets, often benefiting large corporations; many unbanked people	Promotes financial inclusion, fair credit access, and ethical investing (socio- and eco-conscious); microfinance and cooperative banks; DeFi	Centralized banking and credit systems, with state-controlled interest rates, financial flows, and approvals (e.g. social scoring system for loans)
Innovation, entrepreneurship and growth	Innovation is driven by profit incentives, competition, and venture capitalism; GDP growth	Innovation with social impact, ethical business practices, cooperative ownership; inclusive growth	State directs innovation; focus on strategic industries and national goals for growth that serve the state's agenda
Global trade and economic integration	Advocates for free trade and globalization, often favouring MNCs	Supports fair trade and regional economic cooperation without exploitation	Tends to favour protectionism, state-led trade agreements, and regional self-sufficiency
Examples	United States, UK (1980s-2000s), emphasizing deregulation and privatization	Nordic model (e.g., Sweden, Norway) with strong welfare states and inclusive policies; Rwanda, African collectivism	Historical examples: Soviet Union, Maoist China; modern example: Cuba, China's mixed economy

Sources Van Niekerk (2022), Gerstle (2022), Bolesta (2022) it ensures that all individuals and communities, regardless of their background, have access to the resources, opportunities, and benefits needed to thrive. By also promoting social justice, it creates a fairer society where everyone has the chance to succeed. An inclusive economy ensures ecological parity as it emphasizes the fair and sustainable use of natural resources to protect the environment and ensure that its benefits are shared equitably. When economic systems are designed to be inclusive, they can promote ecological parity by addressing the root causes of environmental degradation and ensuring that the benefits of sustainable practices are distributed fairly.

- *The sharing economy and the circular economy*: The sharing and circular economies are two transformative economic models that, when combined, can really accelerate economic inclusion. The sharing economy prioritize shared access to resources, while the circular economy reduces waste and focus on reusing materials. Together, they create opportunities for marginalized groups, reduce inequality, open doors for new business ventures, and foster sustainable growth. They do this through increased access to resources (e.g., transportation, housing, tools), job creation and entrepreneurship (e.g., in recycling and repair), empowering marginalized groups (e.g., flexible work opportunities for women), community building and social cohesion (e.g., local networks), access to technology and innovation (e.g., digital platforms that connect diverse people), and skills development (e.g., learning skills in digital literacy, repair, and sustainable practices).
- *Inclusive policies and governance*: Economic policies should be inclusive, participatory, and accountable to build trust and ensure that resources are distributed fairly. Inclusive policies that align with people's needs and address critical structural changes are crucial for establishing an economy that works for everyone. In an inclusive economy, labour unions, cooperatives, and community groups have a say in economic decision-making. Anti-discrimination laws are important, putting proper legal frameworks in place that prevent discrimination based on gender, race, religion, disability, or other factors. Marginalized communities should be involved in policy-making processes to ensure that their needs are addressed. Monetary policy and the reform of financial systems are important levers for promoting economic and financial inclusion. Intentional and well-designed fiscal policies are necessary to counteract inequalities and build an economy that benefits all.
- *Fair trade and global economic integration*: Ethical trade policies are necessary to support fair trade agreements that promote worker rights and sustainable production. Trade and investment should benefit all sectors of society, not just corporations and elites. The redistribution of export earnings to the workers are vital for economic development. In terms of regional and local development, economic growth should benefit both urban and rural areas, avoiding excessive concentration of wealth. This will empower marginalized producers, create more access to global markets, diversify economies, facilitate technology and skill transfers, and attract FDI, which can be used to build infrastructure, create

jobs, and stimulate economic growth. Global economic integration and globalization goes hand in hand, but the inclusive economy has a unique stance on globalization.

Globalization—in whichever form—is an integral part of the twenty-first century global economy, given the growing interdependencies between economies—though not without growing tension as well. The current trajectory of globalization has been accompanied by increased isolation and exclusion, leading to growing imbalances between the rich and poor, and between developed and underdeveloped nations. The inclusive economy takes a balanced and reformative approach to globalization, recognizing both its benefits and challenges. Rather than rejecting globalization outright, it seeks to restructure it into a more just and sustainable system that prioritizes equity and shared prosperity (Ajide et al., 2021). It is critical of a purely growth-driven, neoliberal approach to globalization.

The inclusive economy promotes a model where global trade and economic integration uplift all communities, rather than concentrating power and wealth at the top. Calling for a fundamental reorientation of its principles and practices, the inclusive economy's stance on globalization is one of critical engagement. It seeks to maximize globalization's benefits—such as economic growth, innovation, and cultural exchange—while addressing its challenges, including inequality, exploitation, and environmental harm (Van Niekerk, 2019). By promoting fair trade, ethical practices, and global cooperation, an inclusive economy aims to create a globalization that works for everyone, leaving no one behind. This approach ensures that globalization becomes a force for shared prosperity, sustainability, and social equity in the twenty-first century.

Moving on, another important question in the context of a new global economy taking shape, is: Why are inclusive economies needed? The short answer is: They are vitally important to ensure economic sustainability in the global economy in order to address the twin crisis of environmental degradation and economic inequality/exclusion. Arguably, only an inclusive economy has the right ingredients to address both crises—but it requires worldwide implementation. The world is already in deep trouble because its ecological footprint is greater than its biocapacity, and 90% of the world's population live on 10% of the earth's resources (while 10% live on 90% of the resources) (Fioramonti, 2017). Both these scenarios are simply unsustainable, setting us up for a major existential crisis. So, what will change in the global economy of the twenty-first century if economies become more inclusive?

It is expected that such a major positive change would lead to significant and fundamental shifts across various aspects of the global economic landscape, such as global economic structures, policies, and business practices. These changes would impact growth patterns, wealth distribution, trade, sustainability, and financial systems—creating a more equitable, resilient, and sustainable global economy. Arguably, the first major effect would be a new, redefined purpose for the economy. The primary goal of the global economy would shift from a sole focus on economic growth (measured by GDP) to achieving collective well-being, genuine economic progress, and sustainability for both people and the planet. The economy would be seen as a tool to organize society for the distribution of tasks and resources to achieve such collective well-being (Van Niekerk, 2022). Economic activities would be increasingly oriented towards prioritizing the needs of people and the health of the planet, alongside preventing global economic crises.

This would involve a move towards low-impact production processes, reduced waste, a green global economy, shared prosperity, fairness, and sustainability. There would be a shift from exclusion to inclusion. The current trend of rising inequality and exclusion would be countered by intentionally reorganizing society on the basis of shared collectivist ideals and ensuring broader participation in the economy (Fioramonti, 2017). Economic inclusion would become a fundamental condition for a sustainable and diversified economy (without any centralization of economic power).

Secondly, the linear, high-wastage economy would gradually transition towards circular economic models that preserve resources, eliminate waste, and regenerate natural systems. In this way, circular economy models, carbon–neutral industries, and green finance would drive innovation and ensure long-term sustainability. Simultaneously, collaborative economic models, emphasizing access over ownership, shared resources and decentralized networks would become more prominent (Mishra et al., 2021). Instead of solely relying on redistribution after periods of growth, inclusive economies would emphasize predistribution through just tax-transfer systems and labour market reforms that help workers earn higher incomes directly. Fair trade principles could also be more widely integrated to ensure justifiable returns for producers. Equitable trade practices will ensure that developing countries benefit from global economic integration, which will reduce global disparities. Businesses would increasingly adopt inclusive business models that prioritize human dignity, rights, economic sustainability, and environmental inclusiveness—while increasing profitability (Doherty & Kittipanya-Ngam, 2021). Advancing conscious capitalism, the focus of business would gradually shift from solely maximizing shareholder value to creating long-term value for all stakeholders. As stated by Roche et al., (2017:18), "we may now, in fact, have a once-in-a-lifetime opportunity to reposition business as a restorative power for healing the global economy; an engine for profound positive change for the many".

Thirdly, the dominance of self-interest would be challenged by a new economic ethos grounded in shared values, such as trust, compassion, collaboration, and a focus on the common good. Principles like Ubuntu, emphasizing interconnectedness, collective well-being, and accountability, would play a more significant role (Moleka, 2025). Given the transparency it will bring through mass data observation by means of the latest technology, this will bring about a fundamental shift in mindset and economic thinking, resetting economic minds away from purely materialistic and short-term perspectives towards a more balanced, interdependent, and long-term view. This would involve embracing new economic thinking that challenges the status quo and promotes a holistic understanding of economic activity. Addressing inequality and exclusion with such a mindset will reduce social tensions and foster trust and cooperation within and between communities. Following it up with investments in underserved areas will improve infrastructure, create jobs, and enhance quality of life, which will strengthen social bonds and support networks.

Fourthly, more opportunities will be created to leverage technology for inclusion. Tech innovation often benefits corporate profits over societal needs. This exacerbates inequalities. In an inclusive economy, AI, blockchain, and digital platforms are leveraged for inclusive finance, e-learning, and healthcare access to bridge the digital divide. Innovation and technological advancements would be intentionally directed towards improving distribution, creating new access, and enhancing circular economic processes, rather than primarily focusing on optimizing production in ways that lead to exclusion. One area that will benefit significantly is financial inclusion. Microfinance, digital banking, and decentralized finance (DeFi) will provide increasing access to financial services, empowering individuals and businesses worldwide (Ozili, 2022). Inclusive innovation would also lead to increased productivity. By including marginalized groups, economies will tap into a wider range of skills, perspectives, and ideas, driving innovation and creativity. This will build support for small businesses and start-ups, particularly those owned by underrepresented groups, which will spur economic dynamism and job creation.

Fifthly, governments and institutions would prioritize the formulation and implementation of inclusive economic policies that are in sync with the needs of the people and promote economic inclusion as a primary goal. This would require institutional reform to embed a culture of inclusivity across the state, market, science, technology, and education. With productive economic inclusion as a shared goal, stakeholders from communities, businesses, civil and non-governmental organisations (NGOs), governments, and international institutions can work together more efficiently to reorganise society from ground level to the highest level to bring alignment with holistic, inclusive progress (Ahi et al., 2023). This would lead to greater global cooperation and economic diplomacy. Collaboration between nations will spread best practices for inclusive and sustainable development, fostering global equity. One of the key outcomes of this will be a reduction in income and wealth inequality. Progressive taxation, living wages, and wealth redistribution policies—coordinated in the international community—will ensure that economic benefits are widely shared, improving economic equity and fostering social stability. Marginalized groups, including women, minorities, and rural populations, will gain access to education, healthcare, and financial services, thus levelling the playing field. All ethnic, racial, and religious minorities will benefit from anti-discrimination policies and targeted support, which will enable their full participation in the economy. By empowering marginalized groups, inclusive economies will lift millions out of poverty and create a more stable global economic environment.

Sixth, new measures of progress would be implemented. The reliance on GDP as the primary indicator of economic success would diminish (but not disappear), with a greater emphasis on qualitative indicators that measure genuine economic progress, such as the net effect of subtracting negative externalities from GDP and adding positive externalities not measured by GDP. Resilient and sustainable growth will be prioritized. Indicators of well-being, social capital, broad-based prosperity, and environmental sustainability would gain prominence. The Human Development Index (HDI) is a metric that measures these indicators to a certain degree, and while it has some value, on its own it is not sufficient when major policy decisions need to be made.

Another, more comprehensive metric that should be implemented by countries is the Genuine Progress Indicator (GPI). While GDP mainly measures economic activity (transactions), the GPI takes into account a wider range of economic, social, and environmental factors to provide a more holistic snapshot of societal well-being (Sinha, 2023). By providing monetized estimates of societal well-being, the GPI aims to measure the real increase in economic welfare and/or improvement in the well-being of a nation or community as a whole. The GPI starts with personal consumption expenditures (a major component of GDP) and then makes adjustments—adding factors that contribute to well-being but are not included in GDP, and subtracting factors that diminish well-being. The relationship between GDP and GPI is compared to the relationship between a company's gross profit and its net profit, where GPI (net progress) is GDP minus social and environmental costs. In addition, green economic growth will prioritize sustainable practices, such as renewable energy, circular economies, and conservation, reducing environmental degradation and promoting long-term growth. This will be part of the shift from GDP-centric growth to well-being-oriented growth.

Lastly, there would be an enhanced resilience to various types of economic crises. In inclusive economies, much emphasis will be placed on collaboration, decentralized systems, and social capital, making them more resistant to economic shocks, stresses, and financial crises. Research indicates that economic shocks (e.g., financial crises and pandemics) harm the poor and vulnerable disproportionately (Asare & Barfi, 2021). By focusing on preventing social exclusion and regulating capital flows, it would contribute to greater economic stability. Universal social protections, living wages, and diversified supply chains will make economies more enduring to global crises. As a result, inclusive economies will be less reliant on single sectors, making them more resilient to shocks like health crises, natural disasters, or financial crises. This will further bring about a more durable and crisis-proof global economy.

In essence, the establishment of inclusive economies would usher in a paradigm shift in the global economy of the twenty-first century, moving away from a system that often prioritizes profit and growth at the expense of people and the planet towards one that is more equitable, sustainable, resilient, and ultimately focused on enhancing the well-being of all. This transition would require intentional effort, collaboration among all stakeholders, and a fundamental rethinking of economic principles and practices. Although there will be challenges of implementation due to structural and systemic rigidities, inclusive economies will create a world where everyone has the opportunity to thrive. If inclusive economies become the global norm, the world economy would shift from extraction and profit-maximization to a model centered on shared prosperity, sustainability, and resilience. This transformation would foster a stronger middle class, reduce poverty, enhance environmental protection, and create a fairer global system for all. This transition will not only address the sustainability challenges of today's global economy, but also build a foundation for a more just and prosperous future for generations to come.

4 Agenda 2030: Fast-Tracking Progress Towards the SDGs

With less than a decade remaining to achieve the UN Sustainable Development Goals (SDGs), urgent, coordinated, and transformative action is needed to accelerate progress. The world needs bold policies, sustainable financing, digital innovation, and global cooperation. The factors required for the acceleration are deeply intertwined with the principles of an inclusive economy. Achieving the SDGs by 2030 is still possible, but requires urgent, coordinated, and inclusive action from governments, businesses, and civil society. Importantly, strong leadership and adequate financing will largely determine how effective the acceleration is. The following critical factors can fast-track SDG implementation and ensure meaningful impact by 2030:

- *Strong political will and governance*: Commitment from governments at all levels is fundamental for prioritizing and investing in SDG achievement. They must prioritize SDG integration into national policies, budgets, and legal frameworks. It can no longer be an add-on priority, but be intrinsic to policies and national and local development plans (Pattberg & Bäckstrand, 2023). Multilateral cooperation through enhanced global partnerships and regional collaboration are needed to support and drive coordinated action. It will certainly require a team effort. Apart from the results by 2030, the process of collaboration in itself will bring the global community closer together. Accountability mechanisms are as important, establishing transparent monitoring and evaluation systems to track progress and hold stakeholders accountable to ensure that resources are used effectively.
- *Increased financing and SDG investments*: It is vital that public and private financing and adequate resource mobilization for SDG-related projects scale up, especially in developing countries. Governments, businesses, and financial institutions must allocate more resources to SDG projects. States can broaden the tax base, increasing tax progressivity, phase out untargeted and harmful subsidies (e.g., those on fossil fuels and some agricultural commodities) to generate substantial fiscal space for health and other SDG-related investments, and rationalize tax incentives and simplify procedures for ease of compliance to improve revenue volumes (Pasara & Mhlanga, 2023). From the private sector's side, innovative financing can be used by leveraging tools like green bonds, social impact bonds, targeted investments in key sectors, and blended finance to attract ESG investment. By expanding sustainable finance, impact investing, and carbon markets, large-scale funding can be mobilized (Khajuria et al., 2022). The

international community can assist through debt relief, sovereign SDG bonds, debt-for-climate swaps, increased ODA, and fair-trade policies. Reducing the financial burden on especially developing countries can enable them to focus on SDG progress.

- *Digital transformation and innovation*: AI and big data are increasingly being used for SDG tracking, involving data-driven insights to help monitor progress, identify gaps, and optimize resource allocation and improve policy decisions. FinTech and digital inclusion are expanding mobile banking and blockchain, while digital payments improve financial access for the unbanked. Smart infrastructure and renewable tech, such as innovations in clean energy, sustainable agriculture, and water management are accelerating impact. Another example is the use of Digital Public Technologies (DPTs). DPTs are digital assets that create a level playing field for broad access or use through public ownership, regulation, or open-source principles (Ingram et al., 2022). This includes digital public goods and digital public infrastructure (e.g., internet access and platforms). Specific DPT platforms, such as those for personal identification and registration, payments, knowledge sharing, environmental monitoring, and public service delivery, can directly or indirectly contribute to a wide range of SDG outcomes, and address 'internet poverty' (access and affordability). For instance, digital identity systems can ensure access to birth certificates (SDG 16.9), land titles (SDG 1.4), bank accounts (SDG 8.10), social protection (SDG 1.3), schools (SDG 4.1), and health clinics (SDG 3.8) (Abdullah, 2024). Knowledge infrastructure, like India's DIKSHA platform, can provide open educational resources (SDG 4) and support professional skill development across sectors.
- *Strengthening local and grassroots participation*: Local organizations must have a stronger role in SDG implementation to ensure community-led development and to foster ownership and accountability. Increasing public participation (civic engagement) accelerates social and policy change due to increased awareness. Women and the youth, in particular, should be empowered through gender equity and youth leadership, which are critical for long-term sustainability (Huan et al., 2021). Localizing the SDGs is vital; this involves customizing policies to local conditions and needs, and incorporating bottom-up approaches from local communities, to ensure that interventions are relevant and effective. Local communities often possess valuable indigenous knowledge and understanding of their local context, which can be crucial for developing effective and sustainable solutions. Integrating this local knowledge into Science, Technology, and Innovation (STI) efforts can unlock transformational capacity and ensure that solutions are grounded in local realities. By engaging local authorities and communities in national, sustainable development objectives can be invaluable in bridging institutional gaps, and ensuring that local needs are considered in national planning.
- *Global and local partnerships*: Multi-stakeholder collaboration is critical. Transformative partnerships between states, businesses, civil society, and international organizations must constantly be strengthened to pool resources, share

knowledge and technology, and ensure that efforts are relevant and effective at all levels. South-South cooperation among developing countries is vital to address common challenges. The involvement of global institutions at local level offers opportunities and incentivizes local businesses to align their operations and investments with the SDGs through CSR and sustainable business practices (e.g., ESG investing). Multilateral development banks play a crucial role in financing development projects, engaging steadily in collaborative ventures with other stakeholders, including the private sector. International organizations and UN agencies facilitate global action by setting norms, providing technical assistance, and fostering collaboration among member states. Initiatives like the UK Research and Innovation Global Challenges Research Fund Accelerate Hub exemplify shared research initiatives across UN agencies, the African Union, and NGOs to find accelerator synergies for the SDGs. Think tanks and academic institutions, through global networks and partnerships, can generate, translate, and disseminate knowledge relevant to SDG implementation and hold governments and multilateral entities accountable. They can also help bridge the North–South divide by enabling more South-South partnerships. Academia and universities, with their local presence, can be valuable assets in SDG implementation by providing know-how, technical competence, and familiarity with local communities.

- *Sustainable economic growth and job creation*: Two highly effective ways of fast-tracking progress towards achieving the SDGs are generating new employment opportunities and inclusive growth. Results like these build confidence and keep all stakeholders committed to progress and achieving the other goals. Although the full effects of sustainable growth and job creation are only seen over time, immediate action steps can be taken, such as resource efficiency (for SDG 13), waste reduction (SDG 12), sustainable business models (SDG 11), fair wages, inclusive industrialization (SDG 9), gender equality (SDG 5), and job security. These will already contribute to SDG success and when they are reinforced by impact-driven businesses and community-led organisations who are committed to bring change, social challenges and hurdles will be addressed much faster. Sustainable growth and job creation (SDG 8) are powerful enablers and a direct pathway to reducing poverty (SDG 1) and inequality (SDG 10). Increased income enables individuals and families to afford basic necessities like food, clean water, sanitation, healthcare, and education, which are directly linked to SDG 2, SDG 3, and SDG 6. Immediate progress in almost two-thirds of the 17 SDGs are made just through inclusive growth and new jobs. When further human capital development is done in a community after such progress has been started, it activates a positive cycle of change because a skilled and healthy workforce is more productive, contributing further to economic growth and the attainment of other SDGs.
- *Other critical factors that can fast-track SDG progress*: To prevent repetition, there are a number of other factors that have already been discussed elsewhere, but is relevant here. They include climate action and environmental sustainability, global health and social protection systems, education and awareness

(SDG literacy), infrastructure and urban development, conflict resolution and peacebuilding, and cultural and behavioural change.

Fast-tracking SDG progress requires a holistic and integrated approach that addresses the interconnected nature of the 17 SDGs and their 169 targets. Progress on the SDGs requires moving beyond narrow, sectoral approaches and taking advantage of positive synergies between different goals and targets. Systems thinking is essential for making mutually reinforcing progress across the economic, social, and environmental dimensions of sustainable development. Identifying and utilizing leverage points and addressing bottlenecks within the system of SDG targets can accelerate progress (Allen & Malekpour, 2023). Leverage analysis can help determine the most effective points for intervention that yield progress across multiple goals. Identifying and addressing systemic bottlenecks can prevent limitations to overall progress, no matter how significantly other parts of the system may change. Ultimately, as pointed out above, by focusing on strong leadership, adequate financing, innovation, partnerships, inclusivity, and sustainability, the global community can accelerate progress toward Agenda 2030. This will not only help achieve the SDGs but also create a more equitable, resilient, and sustainable world for future generations.

5 East and West: A Shared Vision for the Global Economy

A great transition is necessary. Beyond geopolitical tensions and tech wars between East and West, a shared vision of economic sustainability and genuine progress is what countries need to (and want to) work towards. Although many different paths can be identified, there are significant synergy points and common interest aspects that should propel the world towards achieving this. It is not just about putting long-term goals ahead of short-term gains, it is about asking the question of what kind of world are we creating? On a daily basis, every citizen, government, business and organization have a choice in this. If we don't realize that our collective future is more important than our individual benefits or losses, we won't have a future. So, it is not a question of if, but when. What would be the sensible future of the world that almost every global citizen, as part of the global village, would like to see? If we start with *oikonomos* and treat the members of the global village community as 'family members', then we have a strong point of departure to 'get our house in order'. As the first order of business, the economy needs to transition from a growth-driven model (both neoliberal and socialist) towards a more holistic and sustainable approach centered on inclusivity and well-being. This is not to say that growth is unimportant—it is not—but the economy needs to reconsider its current trajectory at a fundamental level. Growth alone is not enough. A multi-dimensional approach is unavoidable, but it might not currently be the politically desirable one. Going from crisis to crisis could be just as unavoidable if the economy continues on this path and keep making 'growth at any cost' the ultimate aim.

As Milton Friedman (1964) stated, "When that crisis occurs, the actions that are taken depend on the ideas that are lying around. That, I believe, is our basic function: to develop alternatives to existing policies, to keep them alive and available until the politically impossible becomes the politically inevitable".

To align the global economy with the SDGs and Agenda 2030, fundamental changes are needed in economic structures, trade policies, financial systems, and sustainability approaches. The global economy must adopt a transformative and inclusive approach that addresses the interconnected challenges of inequality, environmental degradation, and unsustainable growth. For it to not only survive but thrive, the future global economy must be inclusive, sustainable, resilient, and innovation-driven to ensure shared prosperity for all. To guide economies around the world in making such a transition, a multifaceted approach is needed that should at least include the following key directions:

- *Embracing economic inclusivity as a guiding principle*: The most critical element of the way forward is to prioritize economic inclusivity in all aspects of economic life and to address wealth and income inequality (exclusion) with new vigour. This means designing policies, business practices, and social behaviours with the aim of including all members of society in the growth process and ensuring that they benefit from economic progress. This requires a conscious effort to move from exclusion to inclusion, reorganizing society based on shared collectivist ideals, as exemplified by Ubuntu.
- *Redefining progress beyond GDP growth*: The global economy needs to move beyond the narrow focus on GDP as the primary measure of economic progress. Instead, the focus should shift—not without GDP—to genuine economic progress indicators that are more comprehensive, such as GPI, HDI and others—all of which urgently need further refining. Prioritizing overall human well-being as the new end goal is essential. This necessitates adopting and integrating these broader measures of progress into policy-making and development plans to provide a more accurate measurement of societal advancement.
- *Reshaping global trade for fair and sustainable growth*: The trade war needs to end, but more than that, trade must become a catalyst through which to ensure labour rights, fair wages, and sustainable sourcing of raw materials to establish ethical and green supply chains. Moving away from over-reliance on a few global suppliers will create diversified, crisis-resistant supply chains. It is vital that trade agreements must prioritize social and environmental considerations, rather than only economic gains. In this way, global commerce can benefit all economies, especially developing nations.
- *Accelerating green growth and the circular economy*: As part of decarbonization and net-zero policies, expanding renewable energy, electric mobility, and carbon pricing is key to mitigating climate change. While some would argue that there are more 'immediate' crises like a potential nuclear crisis, which could be sparked in the Middle East, the climate crisis represents the alarming need for a sustainable global economy in all its dimensions. Reducing waste, recycling materials, and promoting sustainable manufacturing will drive resource

efficiency and create new business and job opportunities. The poor is dying as we speak because of this—among other things—not materializing rapidly enough. Public and private funding through green finance and ESG investments must prioritize climate-conscious projects. By minimizing waste and maximizing value through innovation, the circular economy offers a highly feasible path to a new global economy.

- *Steering the technological revolution and innovation*: Following an approach of 'Tech for sustainable development', AI, blockchain, green tech, IoT and big data can improve governance, financial inclusion, environmental monitoring, and access to education, healthcare, and financial services, particularly in underserved areas. Efforts to expand digital infrastructure should intensify, given that universal internet access is essential to bridge the digital divide and create new economic opportunities (Karanicolas, 2014). Smart manufacturing and Industry 4.0 must be aligned with the principles of sustainability to ensure that automation and AI-driven industrial processes prioritize job creation alongside efficiency. Governments and businesses have a unique opportunity to harness digital innovation to create more accessible, transparent, and sustainable economic systems.
- *Reinventing global financial systems*: Reform the IMF, World Bank and WTO to be more representative and to shift from debt-based financing toward sustainable development financing. This will create more synergies and opportunities for collaboration between them and their Eastern counterparts, the NDB, AIIB and ADB, to establish a more inclusive and integrated global economic governance framework. A restructuring of global debt is required as developing nations need debt relief mechanisms to free up resources for SDG investment. Using FinTech more broadly in the financial sector should accelerate financial inclusion. There is no doubt that alternative financial systems (e.g., BRICS financial initiatives and CBDCs) will challenge the dominance of the US dollar, with the potential of de-dollarization. However, this situation needs to be managed very carefully.
- *A new economic code/ethos based on shared values*: At the heart of the matter regarding the transition of the economy is the willingness to embrace a new, inclusive mindset. A common economic ethic based on shared values and a unified moral code is crucial for guiding the global economy towards a more sustainable and just future. This ethos should prioritize principles such as stewardship of resources, innovation for increased access, building trust, and fostering a new economic culture that balances simplicity and sophistication. This ethical foundation can be instrumental in reforming capitalism to work for everyone. Conscious capitalism is an example of this, being an economic philosophy that promotes ethical business practices, emphasizing the role of businesses in creating social and environmental value, not just financial profit. It is based on the belief that businesses should serve a greater purpose while fostering prosperity for all stakeholders.

A good, shared vision for the global economy that can bring East and West closer together could be centered on achieving an inclusive, sustainable, and resilient global economy driven by innovation and cooperation, with the SDGs as the common framework. The difference, compared to the recent past, is how intentional countries need to become in achieving the SDGs. Such a vision also involves a gradual repurposing of the economy: repurposing business, investment, trade, technological innovation, and growth, with economic inclusion as the benchmark. In this way, a shared vision must balance mutual prosperity, social equity, ecological parity, and technological progress by making sustainability the central focus, while respecting diverse cultural, political, and economic models. This vision must prioritize collaboration over competition, ensuring a more resilient, fair and inclusive global economy. It should bridge cultural, political, and economic differences while addressing shared challenges like climate change, inequality, and technological disruption. Perhaps, to bring it all together, this shared vision for East and West could be: "An innovative global economy of inclusive growth, sustainable development, and shared prosperity for all". A bold vision unites the world in striving for a better future.

6 Conclusion

The global economy of the twenty-first century is at a pivotal juncture as it is shaped by technological innovation, shifting geopolitical and globalization dynamics, the 4IR, financial transformation, and sustainability imperatives. Fundamental changes are redefining global economic structures, which necessitates a rethink of our economic goals, in line with economic inclusion, financial innovation and sustainable development as key drivers of long-term prosperity. This is shaping a new vision and new potential for the global economy—unlike ever seen. In an era defined by rapid change and uncertainty, FinTech, blockchain, and CBDCs are restructuring financial systems, creating new opportunities for access and inclusion while raising concerns about regulation and stability. Adding to its uniqueness, this era includes the rise of BRICS+, de-dollarization, and economic nationalism, which are challenging traditional Western-led financial institutions and fostering a more multipolar world order. This order, therefore, also present major risks, especially if the geopolitical tension is not navigated carefully. Even a catastrophe like a world war is not excluded.

The disruptive nature of the 4IR on industries and labour markets, the fragmenting effects of geopolitical tensions and the rise of economic nationalism is bringing about a dynamic evolution of globalization towards regionalization and strategic alignments (Olaitan & Mapanga, 2024). By bridging these interconnected themes, this chapter contributes to the broader discourse on how the global economy can transition toward a more inclusive and sustainable model. It reinforces the importance of balancing technological advancement with equitable economic policies, ensuring that growth benefits all—not just a select few. By drawing upon the principles of an inclusive economy and the framework of Agenda 2030, this chapter has substantiated the urgent need to move beyond conventional growth-centric models.

The findings suggest that without proactive governance, international cooperation, and responsible innovation, economic disparities will continue to widen, threatening social stability, long-term economic security, and even global safety due to the threat of large-scale military conflict. The implications of these complex shifts extend far beyond financial markets; they reshape global governance, trade relations, and economic power structures. Nations and institutions must adapt to these transformations by fostering greater multilateral collaboration, ethical business practices, and inclusive economic policies that prioritize the well-being of both people and the planet. Failing to do so risks exacerbating global inequalities, weakening financial resilience, and undermining collective progress toward the SDGs. The transformations underway necessitate proactive policy responses to address job displacement and inequality stemming from the 4IR, to navigate the complexities of a fragmented geoeconomic landscape, and to foster a more equitable and resilient form of globalization. Embracing economic sustainability and prioritizing the SDGs are not merely aspirational goals but fundamental requirements for long-term global stability and prosperity.

One area of particular concern regarding the stability of the global economy and the central role of the United States in it is the growing uncertainty about de-dollarization. Apart from a nuclear disaster, this is the one 'uncontrollable' factor that the new Trump administration cautioned—and even warned—against (Wearden, 2024). To avoid the precarious path of a huge global catastrophe due to de-dollarization, a credible and inclusive dialogue on a new global monetary agreement should begin now. Major economies, including the US, must reach a consensus on a revised monetary system—potentially backed by gold, commodities and/or crypto (Giustra, 2023). This transition would be challenging for the US, potentially to a politically unfeasible degree. Arguably, the most viable approach would be a gradual reduction in global reliance on the dollar over an extended period, allowing both the US and other nations to adapt. This would achieve multiple goals. A multipolar monetary system could also create a more level playing field for poorer countries and potentially foster greater long-term economic and political stability worldwide. While the transition would likely be disruptive and could lead to a decline in the US standard of living, it remains a more manageable alternative to the severe economic upheavals seen in history.

Reflecting on the broader context, the twenty-first century demands a paradigm shift in economic thinking, practice and vision. The interlocked crises of environmental degradation and economic inequality necessitate a *shared global vision* that transcends geopolitical divides. The global economy stands at a defining moment: it demands bold, forward-thinking strategies to ensure shared prosperity, stability, and sustainability. The alternative could be a catastrophic scenario—one of devastation that no one desires and that would be unbearable. A new commitment to collaborate is required from both the East and the West.

As new economic paradigms emerge, the challenge is not just to manage change but to shape it actively in ways that empower people, protect the planet, and create a more just and resilient world. The correct path forward is to embrace inclusive, innovative, and sustainable economic practices—not just as an option, but a necessity for the future of global prosperity. The trajectory of the new global economy hinges on our collective ability to embrace a fundamental reorientation towards inclusivity and sustainability, while circumventing all forms of conflict/war. By intentionally repurposing economic activities, fostering global cooperation, and prioritizing the well-being of all, we can navigate the complexities of this era and build a future global economy characterized by innovative growth, shared prosperity, and lasting resilience for generations to come. We live in an era of extremes that demands extraordinary leadership and unwavering commitment.

References

Abdullah, W. (2024). Artificial intelligence for achieving sustainable development goals: Applications, techniques and progress. *International Journal of Computers and Informatics, 5*, 117–128.

Acharya, A., Estevadeordal, A., & Goodman, L. W. (2019). Reshaping global order in the 21st century: G-plus leadership in a multiplex world. *China & World Economy, 27*(5), 63–78.

Acs, Z. J., Song, A. K., Szerb, L., Audretsch, D. B., & Komlósi, É. (2021). The evolution of the global digital platform economy: 1971–2021. *Small Business Economics, 57*, 1629–1659.

Aggarwal, P. (2020). On de-risking and de-dollarizing intra-BRICS trade via smart contracts. *BRICS Journal of Economics, 1*(4), 54–69.

Aggarwal, V. K., & Reddie, A. W. (2021). Economic statecraft in the 21st century: Implications for the future of the global trade regime. *World Trade Review, 20*(2), 137–151.

Ahi, A. A., Sinkovics, N., & Sinkovics, R. R. (2023). E-commerce policy and the global economy: A path to more inclusive development? *Management International Review, 63*(1), 27–56.

Ajide, F. M., Osinubi, T. T., & Dada, J. T. (2021). Economic globalization, entrepreneurship, and inclusive growth in Africa. *Journal of Economic Integration, 36*(4), 689–717.

Alavijeh, E. E., Kiasejmahaleh, R. R., & Amiri, M. J. (2017). Green economy a way to environmental sustainability. In *International Conference on Agricultural, Natural Resources and Sustainable Environment* (pp. 1–6).

Allen, C., & Malekpour, S. (2023). Unlocking and accelerating transformations to the SDGs: A review of existing knowledge. *Sustainability Science, 18*(4), 1939–1960.

Antràs, P. (2020). De-globalisation? Global value chains in the post-COVID-19 age. *National Bureau of Economic Research*, Working Paper, 28115 (pp. 1–49).

Asare, P., & Barfi, R. (2021). The impact of Covid-19 pandemic on the Global economy: Emphasis on poverty alleviation and economic growth. *Economics, 8*(1), 32–43.

Bolesta, A. (2022). From socialism to capitalism with communist characteristics: The building of a post-socialist developmental state in Central Asia. *Post-Communist Economies, 34*(1), 71–98.

Bolhuis, M. A., Chen, J., & Kett, B. (2023). *The costs of geoeconomic fragmentation*. IMF.

Bosone, C., Dautović, E., Fidora, M., & Stamato, G. (2024). How geopolitics is changing trade. *ECB Economic Bulletin Box, 2*, 1–34.

Buckley, P. J. (2023). Corporate reactions to the fracturing of the global economy. *International Business Review, 32*, 1–9.

Capello, R., & Dellisanti, R. (2024). Regional inequalities in the age of nearshoring. *The World Economy, 47*(10), 4225–4249.

Ciucu, A. N., Jigani, A. I., Sandu, A., & Delcea, C. (2024). Employee happiness and green practices: A case study of internal corporate social responsibility. *International Journal of Behavior Studies in Organizations, 11*, 23–37.

Chen, Y., Zhang, S., & Miao, J. (2023). The negative effects of the US-China trade war on innovation: Evidence from the Chinese ICT industry. *Technovation, 123*, 1–12.

Cricelli, L., & Strazzullo, S. (2021). The economic aspect of digital sustainability: A systematic review. *Sustainability, 13*(15), 1–15.

Cosma, S., & Rimo, G. (2024). Redefining insurance through technology: Achievements and perspectives in Insurtech. *Research in International Business and Finance, 70*, 1–17.

De Koker, L., Morris, N., & Jaffer, S. (2019). Regulating financial services in an era of technological disruption. *Law in Context, 36*(2), 90–112.

Deschacht, N. (2021). The digital revolution and the labour economics of automation: A review. *ROBONOMICS: The Journal of the Automated Economy, 1*, 1–14.

Diesen, G. (2017). The global resurgence of economic nationalism. *Russia in Global Affairs, 16*(2), 1–13.

Doherty, B., & Kittipanya-Ngam, P. (2021). The role of social enterprise hybrid business models in inclusive value chain development. *Sustainability, 13*(2), 1–24.

Erie, M. S., & Streinz, T. (2021). The Beijing effect: China's digital Silk Road and transnational data governance. *Journal of International Law and Politics, 54*(1), 1–92.

Fioramonti, L. (2017). *Wellbeing economy: Success in a world without growth.* Pan Macmillan South Africa.

Friedman, M. (1964). *Capitalism and freedom: With the assistance of Rose D. Friedman.* University of Chicago Press.

Galeone, G., Ranaldo, S., & Fusco, A. (2024). ESG and FinTech: Are they connected? *Research in International Business and Finance, 69*, 1–14.

Gaudiosi, F. (2018). Economic nationalism and the post-global future. *International Affairs Institute, 18*(4), 1–5.

Gerstle, G. (2022). *The rise and fall of the neoliberal order: America and the world in the free market era.* Oxford University Press.

Giustra, F. (2023). *De-dollarization: Not a matter of if, but when.* https://responsiblestatecraft.org/2023/05/03/de-dollarization-not-a-matter-of-if-but-when/

Goldberg, P. K., & Reed, T. (2023). Is the global economy deglobalizing? If so, why? And what is next? *Brookings Papers on Economic Activity, 1*, 347–423.

Gouvea, R., & Gutierrez, M. (2023). "BRICS Plus": A new global economic paradigm in the making? *Modern Economy, 14*(5), 539–550.

Hasan, M., Hoque, A., Abedin, M. Z., & Gasbarro, D. (2024). FinTech and sustainable development: A systematic thematic analysis using human-and machine-generated processing. *International Review of Financial Analysis, 95*, 1–20.

Helleiner, E. (2021). The diversity of economic nationalism. *New Political Economy, 26*(2), 229–238.

Huan, Y., Liang, T., Li, H., & Zhang, C. (2021). A systematic method for assessing progress of achieving sustainable development goals: A case study of 15 countries. *Science of the Total Environment, 752*, 1–24.

Hulten, C. R. (2015). Measuring the economy of the 21st century. *National Bureau of Economic Research (NBER), 4*, 1–7.

Ingram, G., McArthur, J. W., & Vora, P. (2022). How can digital public technologies accelerate progress on the sustainable development goals. *Brookings Global Working Paper, 174*, 1–36.

Jalinous, F., & Sensenig, T. (2022). Keeping up with CFIUS: A practitioner's report on national security trends in the United States. *Erasmus Law Review, 15*(4), 266–277.

Javaheri, D., Fahmideh, M., Chizari, H., Lalbakhsh, P., & Hur, J. (2024). Cybersecurity threats in FinTech: A systematic review. *Expert Systems with Applications, 241*, 1–31.

Javaid, M., Haleem, A., Singh, R. P., & Sinha, A. K. (2024). Digital economy to improve the culture of industry 4.0: A study on features, implementation and challenges. *Green Technologies and Sustainability, 2*, 1–19.

Jędrzejowska, K. (2017). Economic nationalism: Old-new phenomenon in the global economy. *Zeszyty Naukowe Polskiego Towarzystwa Ekonomicznego w Zielonej Górze, 4*(7), 156–167.

Kalu, E. N., & Ogbonnaya, O. E. (2019). Globalization and economic nationalism: Engaging the perspectives. *Journal of Arts and Management, 4*(2), 63–73.

Karanicolas, M. (2014). Bridging the divide: Understanding and implementing access to the Internet as a human right. *The Journal of Community Informatics, 10*(2), 1–8.

Karim, S., & Lucey, B. M. (2024). BigTech, FinTech, and banks: A tangle or unity? *Finance Research Letters, 64*, 1–10.

Kisiel, R., & Lubowiedzka, M. (2010). Human capital in modern economy-importance, methods of measurement and investments. *Olsztyn Economic Journal, 5*(1), 1–14.

Khajuria, A., Atienza, V. A., Chavanich, S., Henning, W., Islam, I., Kral, U., et al. (2022). Accelerating circular economy solutions to achieve the 2030 agenda for sustainable development goals. *Circular Economy, 1*(1), 1–9.

Khan, H. H., Kutan, A. M., & Qureshi, F. (2024). Fintech integration: Driving efficiency in banking institutions across the developing nations. *Finance Research Letters, 67*, 1–11.

Knell, M. (2021). The digital revolution and digitalized network society. *Review of Evolutionary Political Economy, 2*(1), 9–25.

Kraut, R. (2006). *The Blackwell guide to Aristotle's Nicomachean ethics*. Blackwell.

Larionova, M., & Shelepov, A. (2022). BRICS, G20 and global economic governance reform. *International Political Science Review, 43*(4), 512–530.

LaRocco, L.A. (2025). *Trump is targeting China-made containerships in new flank of global economic war on the oceans*. https://www.cnbc.com/2025/03/11/trump-pursues-new-trade-war-on-seas-targeting-china-containerships.html

Liu, Z. Z., & Papa, M. (2022). *Can BRICS de-dollarize the global financial system?* Cambridge University Press.

Mathews, J. A. (2006). Dragon multinationals: New players in 21st century globalization. *Asia Pacific Journal of Management, 23*, 5–27.

Matyushok, V., Krasavina, V., Berezin, A., & García, S. J. (2021). The global economy in technological transformation conditions: A review of modern trends. *Economic Research-Ekonomska Istraživanja, 34*(1), 1471–1497.

Mishra, J. L., Chiwenga, K. D., & Ali, K. (2021). Collaboration as an enabler for circular economy: A case study of a developing country. *Management Decision, 59*(8), 1784–1800.

Mikecz, R. (2019). The cornerstone of economic nationalism: National self-image. *Fudan Journal of the Humanities and Social Sciences, 12*(4), 587–608.

Moleka, P. (2025). Ubuntu and sustainable cities in Africa. In *The Palgrave handbook of Ubuntu, inequality and sustainable development* (pp. 355–370). Springer.

Moch, E. (2024). The economic and geopolitical significance of the BRICS nations: A comparative analysis of their global influence in the 21st century. *East African Journal of Business and Economics, 7*(2), 86–100.

Mohamad, A., & Dimitriou, D. (2024). From scam to heist: The impact of cybercrimes on cryptocurrencies. *Applied Economics Letters*, 1–10.

Mollan, S., & Geesin, B. (2020). Donald Trump and Trumpism: Leadership, ideology and narrative of the business executive turned politician. *Organization, 27*(3), 405–418.

Mosharrafa, R. A. (2024). Navigating the tides of de-dollarization: Impact on global economy and BRICS initiatives. *World Review of Political Economy, 15*(4), 566–581.

Muntaner, C., Lynch, J., & Oates, G. L. (2020). The social class determinants of income inequality and social cohesion. *The Political Economy of Social Inequalities*, 367–399.

Nach, M., & Ncwadi, R. (2024). BRICS economic integration: Prospects and challenges. *South African Journal of International Affairs, 31*(2), 151–166.

Nahm, J. (2021). *Collaborative advantage: Forging green industries in the new global economy*. Oxford University Press.

Nakano, T. (2007). Alfred Marshall's economic nationalism. *Nations and Nationalism, 13*(1), 57–76.

Nieuwenhuijs, J. (2023). *The Shanghai International Gold Exchange and its role in de-dollarization*. https://www.thegoldobserver.com/p/the-shanghai-international-gold-exchange

Nikolaou, I. E., Jones, N., & Stefanakis, A. (2021). Circular economy and sustainability: The past, the present and the future directions. *Circular Economy and Sustainability, 1*, 1–20.

Nyagadza, B., Pashapa, R., Chare, A., Mazuruse, G., & Hove, P. K. (2022). Digital technologies, Fourth Industrial Revolution (4IR) and Global Value Chains (GVCs) nexus with emerging economies' future industrial innovation dynamics. *Cogent Economics & Finance, 10*(1), 1–14.

Obstfeld, M. (2021). Globalization and nationalism: Retrospect and prospect. *Contemporary Economic Policy, 39*(4), 675–690.

Olaitan, O. O., & Mapanga, A. (2024). Unleashing the power of the 4IR in organisational value chains: A conceptual analysis. *South African Journal of Information Management, 26*(1), 1–12.

Ozili, P. K. (2021). Financial inclusion research around the world: A review. *Forum for Social Economics, 50*(4), 457–479.

Ozili, P. K. (2022). Decentralized finance research and developments around the world. *Journal of Banking and Financial Technology, 6*(2), 117–133.

Pang, D., Jin, X., Zheng, K., & Tien, N. H. (2024). A road toward green growth: Optimizing the role of mineral resources, FinTech innovation and effective governance in G-20 economies. *Resources Policy, 92*, 1–9.

Pandey, V., Sircar, A., Bist, N., Solanki, K., & Yadav, K. (2023). Accelerating the renewable energy sector through Industry 4.0: Optimization opportunities in the digital revolution. *International Journal of Innovation Studies, 7*(2), 171–188.

Pasara, M. T., & Mhlanga, D. (2023). Accelerating sustainable development goals in the wake of COVID-19: The role of higher education institutions in South Africa. *Emerald Open Research, 1*(3), 1–16.

Passaris, C. E. (2006). The business of globalization and the globalization of business. *Journal of Comparative International Management, 9*(1), 3–18.

Patel, V., Chesmore, A., Legner, C. M., & Pandey, S. (2022). Trends in workplace wearable technologies and connected-worker solutions for next-generation occupational safety, health, and productivity. *Advanced Intelligent Systems, 4*(1), 1–30.

Pattberg, P., & Bäckstrand, K. (2023). Enhancing the achievement of the SDGs: Lessons learned at the half-way point of the 2030 Agenda. *International Environmental Agreements: Politics, Law and Economics, 23*(2), 107–114.

Pereira, E. G., Koenck, A., Vitto, W. A. C., Delgado, F., Lopes, J. V. C., & Marins de Carvalho, A. C. (2023). Brazil's local content requirements: Evolution, lessons learned and international trade limitations. *Oil and Gas, Natural Resources, and Energy Journal, 9*(3), 385–423.

Pryke, S. (2012). Economic nationalism: Theory, history and prospects. *Global Policy, 3*(3), 281–291.

Rajan, K. (2025). Decoding DeFi: *A beginner's guide to decentralized finance and yield farming*. AMBCrypto, Tallinn, Estonia.

Regnerova, O., Šálková, D., & Šánová, P. (2021). The importance of industry in modern economies of the globalized world in the 21st century. *In SHS Web of Conferences, 92*, 1–10.

Rammal, H. G., Rose, E. L., Ghauri, P. N., Jensen, P. D. Ø., Kipping, M., Petersen, B., & Scerri, M. (2022). Economic nationalism and internationalization of services: Review and research agenda. *Journal of World Business, 57*(3), 1–13.

Raworth, K. (2018). *Doughnut economics: Seven ways to think like a 21st century economist*. Chelsea Green Publishing.

Rewizorski, M. (2021). Between hyperglobalization and national policy. Is there a way to mitigate populist risks for global economic governance in the post-COVID-19 world? *International Organisations Research Journal, 16*(2), 132–156.

Rizvi, S. K. A., Rahat, B., Naqvi, B., & Umar, M. (2024). Revolutionizing finance: The synergy of FinTech, digital adoption, and innovation. *Technological Forecasting and Social Change, 200*, 1–14.

Roche, B., Jakub, J., Mayer, C., & Radvan, M. (2017). *Completing capitalism: Heal business to heal the world*. Berrett-Koehler Publishers.

Rosário, A. T., & Dias, J. C. (2023). The new digital economy and sustainability: Challenges and opportunities. *Sustainability, 15*(14), 1–23.

Sant'Anna, D. M., & Figueiredo, P. N. (2024). FinTech innovation: Is it beneficial or detrimental to financial inclusion and financial stability? A systematic literature review and research directions. *Emerging Markets Review, 60*, 1–21.

Scott, A. J. (2011). A world in emergence: Notes toward a resynthesis of urban-economic geography for the 21st century. *Urban Geography, 32*(6), 845–870.

Schoneveld, G. C. (2020). Sustainable business models for inclusive growth: Towards a conceptual foundation of inclusive business. *Journal of Cleaner Production, 277*, 1–13.

Shameem, C. C., & Jayaprasad, K. (2020). The evolution of BRICS in international political economy. *American Review of Political Economy, 15*(1), 1–19.

Sinha, A.R. (2023). Genuine progress indicator (GPI): A real indicator of economic progress. *PURUSHARTHA—A Journal of Management, Ethics and Spirituality, 16*(1), 78–92.

Skare, M., & Soriano, D. R. (2021). How globalization is changing digital technology adoption: An international perspective. *Journal of Innovation & Knowledge, 6*(4), 222–233.

Staiger, R. W. (2022). *A world trading system for the 21st century*. MIT Press.

Tarawneh, A., Abdul-Rahman, A., Mohd Amin, S. I., & Ghazali, M. F. (2024). A systematic review of FinTech and banking profitability. *International Journal of Financial Studies, 12*(1), 1–21.

Trushkina, N. (2022). Green economy in the conditions of modern challenges: Conceptual frameworks. *International Science Journal of Management, Economics & Finance, 1*(1), 1–6.

Tung, R. L., Zander, I., & Fang, T. (2023). The tech Cold War, the multipolarization of the world economy, and IB research. *International Business Review, 32*(6), 1–14.

Van Niekerk, A. J. (2019). A conceptual framework for inclusive economics. *South African Journal of Economic and Management Sciences, 22*(1), 1–9.

Van Niekerk, A. J. (2020). Inclusive economic sustainability: SDGs and global inequality. *Sustainability, 12*(13), 1–19.

Van Niekerk, A. J. (2024). Economic inclusion: Green finance and the SDGs. *Sustainability, 16*(3), 1–21.

Van Niekerk, A. J. (2022). *The Inclusive Economy: Criteria, Principles and Ubuntu*. UJ Press.

Vyas-Doorgapersad, S. (2022). The role of BRICS in global governance to promote economic development. *Africa's Public Service Delivery and Performance Review, 10*(1), 1–9.

Walker, G. (2024). *BRICS has high hopes for a new economic world order*. https://intpolicydigest.org/brics-has-high-hopes-for-a-new-economic-world-order/

Wearden, G. (2024). Trump threat of 100% tariffs against Brics nations raises trade war fears. *The Guardian*. https://www.theguardian.com/business/2024/dec/01/trump-threat-100-per-cent-tariffs-brics-nations-dollar

The manufacturer's authorised representative in the EU is Springer Nature Customer Service Centre GmbH, Europaplatz 3, 69115 Heidelberg, Germany. If you have any concerns regarding our products, please contact ProductSafety@springernature.com

Printed and bound by CPI Group (UK) Ltd, Croydon, CR0 4YY
05/10/2026
02228130-0002